Information Dump

I must thank WSB(Atlanta), The Atlanta Journal-Constitution. The daily citizen news (Dalton) and FOX 5(Atlanta). These four news agencies refused to run a story about the crimes documented in this book and thus created this book. Thank you so much for showing why many people view legacy journalism as a joke. Except for editors, good editors are a treasure.

Theory

The US Army and by extension the Georgia Army National Guard have core values. They are as follows:

Loyalty

The first of the Army Values. Bear true faith and allegiance to the U.S. Constitution, the Army, your unit, and other Soldiers. Bearing true faith and allegiance is a matter of believing in and devoting yourself to something or someone. A loyal Soldier is one who supports the leadership and stands up for fellow Soldiers. By wearing the uniform of the U.S. Army, you are expressing your loyalty. And by doing your share, you show your loyalty to your unit.

By this metric I would imagine that if the situation arose a member of the army is expected to be loyal to the law and stand up for a fellow soldier hat is being attacked.

Duty

The second of the Army Values. Fulfill your obligations. Doing your duty means more than carrying out your assigned tasks. Duty means being able to accomplish tasks as part of a team. The work of the U.S. Army is a complex combination of missions, tasks, and responsibilities — all in constant motion. Our work entails building one assignment onto another. You fulfill your obligations as a part of your unit every time you resist the temptation to take "shortcuts" that might undermine the integrity of the final product.

So, this would suggest that if a situation presented itself wherein a shortcut was pushed upon a soldier the soldier would not take the shortcut. If an important paper is lost, then get everyone that signed it back and have them sign the replacement paper.

Respect

The third of the Army Values. Treat people as they should be treated. In the Soldier's Code, we pledge to "treat others with dignity and respect while expecting others to do the same." Respect is what allows us to appreciate the best in other people. Respect is trusting that all people have done their jobs and fulfilled their duty. And self-respect is a vital ingredient with the Army value of respect, which results from knowing you have put forth your best effort. The Army is one team and each of us has something to contribute.

I would think that respect means that a person in charge of a convoy would not just leave a soldier stranded in a combat zone. Even if the first sergeant does not personally like that soldier.

Selfless Service

The fourth of the Army Values. Put the welfare of the Nation, the Army, and your subordinates before your own. Selfless service is larger than just one person. In serving your country, you are doing your duty loyally without thought of recognition or gain. The basic building block of selfless service is the commitment of each team member to go a little further, endure a little longer, and look a little closer to see how he or she can add to the effort.

If a soldier needed you help and you could help even if it took effort, you should.

Honor

The fifth of the Army Values. Live up to Army values. The Nation's highest military award is The Medal of Honor. This award goes to Soldiers who make honor a matter of daily living — Soldiers who develop the habit of being honorable and solidify that habit with every value choice they make. Honor is a matter of carrying out, acting, and living the values of respect, duty, loyalty, selfless service, integrity, and personal courage in everything you do.

Keep your word. Don't lie.

Integrity

The sixth of the Army Values. Do what is right, legally, and morally. Integrity is a quality you develop by adhering to moral principles. It requires that you do and say nothing that deceives others. As your integrity grows, so does the trust others place in you. The more choices you make based on integrity, the more this highly prized value will affect your relationships with family and friends, and, finally, the fundamental acceptance of yourself.

Don't falsify documents and signature, don't protect people that forge documents and signature, don't attack people that report crime.

Personal Courage

The seventh of the Army Values. Face fear, danger, or adversity (physical or moral). Personal courage has long been associated with our Army. With physical courage, it is a matter of enduring physical duress and at times risking personal safety. Facing moral fear or adversity may be a long, slow process of continuing forward on the right path, especially if taking those actions is not popular with others. You can build your personal courage by daily standing up for and acting upon the things that you know are honorable.

Don't be afraid to tell your golf buddy that he is allowing and committing multiple crimes to take place.

Those are directly from the Army's website except for my comments. Obviously, that means they are very important. Let us see them in action. By that I mean what would happen if a Georgia Army National Guard unit and eventually the Georgia National Guard had crimes committed by high-ranking people. Would the core values be followed, or would they be ignored?

Breakaway

Here is today's personal story. There are many good reasons why folks do not report sex discrimination, sexual harassment and sexual assault. I was a victim of sexual harassment while augmenting Security Forces at Bolling AFB, D.C. and there are often times that I wonder if reporting it was worth the cost. "The cost" comes in many forms. For me, I was placed on a detail where I was alone for 12 hours a day. I was removed from having any interaction with anyone at all. I was blamed for having their ranking NCO removed. When I returned from that assignment and went back to the Honor Guard, I went from being one of the top members of Firing Party to being hidden away in the Armory until my time in the Honor Guard was done. When it came time for me to retrain into another field I wanted to apply for OSI (Office of Special Investigations) but my Commander made it clear that she would not endorse that move. Once I did cross train into another field (not OSI), it became really clear to me that my brush with sexual harassment had followed me. I was often made to feel like I was "being put in my place" or "being taught a lesson". When I finally stood up for myself I was slapped with an Article 15 and had a strip taken from me. Even today I don't even like the graduate degree I completed. I am in a career limbo and often find myself wondering how things would have been different had I just brushed off the harassment as others had before me. Maybe I would have gotten accepted into OSI. Maybe I would have stayed in the military and became an officer until retirement. So for those that have ever wondered why people do not report these things or why it can take many years before they come forward, it's because most of the time is doesn't make your life better to do so. I am still feeling the repercussions of what happened to me 20 years ago.

*And I never told anyone in my family or friends at home what I was dealing with because I didn't want to put that burden on them. I didn't tell the few friends I had in the military either. And you can imagine how EVERYONE turned against me after the article 15 and losing rank. I literally lived in almost complete solitude for my last year in the military. My civilian neighbors were the only thing that kept me sane. Thank you Lin Sandmark Harris *

That did not happen to me it happened to a friend of mine. Someone who was dedicated to the military and always did a little extra immediately thrown under the bus to protect a criminal of rank. Is that the rule or the exception? The pain, the sense of betrayal, was that normal or just a result of things that just happened to happen? That phrase is quite common in the military. I have to say that what happened to her angers me more than what happened to me, why is sexual harassment and assault still allowed and even encouraged in the military? This has very little to do with most of the book but how did the hash tag me to and hash tag times up movements miss the military? In the ten years I was in I saw the following:

1. Massive amounts of infidelity, people take their wedding rings off when the plane takes off. People both on and off the plane.

2. People in leadership positions arranging for people to be deployed because they wanted the deployed person's spouse or partner in a sexual way. Wasn't that in the Bible? Or Torah? Or Koran? Or almost every religion.

3. Sexual crimes being swept under the rug or under punished due to high rank.

4. Sexual crimes being over punished due to low rank.

5. Out of control aggressive behavior. If the rank is high, you're not a stalker.

6. Being asked to lie for a supervisor over marital infidelity and drunk on duty multiple times.

7. Use of prostitutes.

8. Use of rank to convince sexual assault victims to perjure themselves or to lie.

9. Massive amounts of sexploitation. (I made that word up)

10. Service members with a ready-made sob story about how they have never cheated on their spouse but whoever they are talking to right then has a special spark within them.

11. I to was threatened and given the run around when I put in for OSI. OSI also handles Air Force internal affairs. Why would you threaten a person putting in for internal affairs?

Everyone sees these things, but few people bring them up. There is intense pressure to either look the other way or join in. Integrity is thrown out the window. Kind of like dealing with members of the Georgia National Guard who have no problem telling lies to anyone and everyone. There is no shame because shame involves too much thought.

Some of the emails and documents from the Georgia National Guard in which Georgia National Guard members say everything and nothing sound remarkably familiar when looking at what happened to my friend. It's almost like they were following a script of how to attack people that report crime. If only there was some law about that. Basically, it comes down to being made an outcast if one reports crime when the criminal is one of rank or status. If someone does, they will be treated as the criminal while the criminal is treated as the victim. So here we go.

Chapter One: Who signed this?

Imagine finding yourself in a situation that seems to be out of a movie or suspense book. Things that you think are fiction suddenly become reality. Document forgery, false charges, signature forgery, whistle blower retaliation, all these things are common and accepted by a military unit. What is going on? It was like a movie with the exception being that there are no end credits, someone might think there are end credits but that is not the case. Just because the person reporting the crimes has been pushed out of the little club doesn't mean the game is over. Maybe the game just got moved to a level playing field. But where did it all start?

This all started at Fort Mcclellan Alabama. The documents that follow show two entirely different things however that is fine. Nobody in the 1-108th knew what was going on according to the 1-108th. But they did know that if anything illegal ever happened, it was just a big mistake that only happened once ever. There was no criminal activity it just happened to happen. This mantra would be repeated far too many times while I was in the Georgia National Guard and even after I was out.

STATE OF GEORGIA
DEPARTMENT OF DEFENSE, MILITARY DIVISION
OFFICE OF THE ADJUTANT GENERAL
PO BOX 17965, ATLANTA, GEORGIA 30316-0965

ORDERS 177-277 26 June 2009

Following order is amended as indicated.

So much of: Orders 009106 this Hq dtd 09 January 2009

Pertaining to:

MOB for MAHAN WESLEY S [redacted] HHC(-), 1ST BN, 108TH AR
(PC2T0-331) PO BOX 36 CALHOUN GA 30701

As reads: ENL TVL/PD: 2192020.0000 B1 BITC 135197 21T1/21T2/22NL/2578/
2579 (SDN) VFRE F4209
ENL TVL/PD: 2192010.0000 B1 BITC 135197 21T1/21T2/22NL/2578/
2579 (SDN) VFRE
How changed: ENL TVL/PD: 2192020.0000 B1 BITC 135197 21T1/21T2
MAH7811TJ9DC74 VFRE F4822 AZUM2E 12161
ENL TVL/PD: 2102020.0000 B1 BITC 135197 21T1/21T2
MAH7811TJ9DC74 VFRE F4822 AZUM2E 12161
Auth: 10 USC 12302a NTE 24 months in mobilized status
APC DJMS_RC: B2EMOB PM/TDC: MOB MOB
Format: 700

FOR THE ADJUTANT GENERAL:

\\\\\\\\\\\\\\\\\\\\\\\\\\\\///////////////
\\ HQ, GAARNG //
\\ OFFICIAL //
\\\\\\\\\\\\\\\\\\\\\\\\\\\\///////////////

DISTRIBUTION:
1-Unit 1-Bn HQ 1-MACOM
1-DCS-FA 1-Individual

THOMAS M. CARDEN JR.
COL, IN, JFHQ-GA
G1

That was one of the forms that was given to me showing that my MOS had been switched from 11 B to 25Q. I was assured everything was ship shape. How little did I know.

ENLISTED PROMOTION POINT WORKSHEET
(For use of this form, see NGR 600-200. The proponent agency is NGB-ARH-S)

Section I - PERSONNEL SYSTEM DATA

1) Name MAHAN WESLEY SCOTT 2) SSN [redacted] 3) RANK SPC
4) Unit name HHC 1ST BN 108TH RSTA (FWD) 5) PMOSD 11B10 6) PULHES 121111
7) DOB 19750403 8) DOR 20061116 [10] 9) PEBD 20001003 [24] 10) BESD 20020627
11) ETS 20511224 12) Active Status Program Z 13) Tech/Selective Service code N

Section II - AWARDS

Fields 14-23 (If the following area is blank, no awards are recorded in the SIDPERS Database)
14) MILDECMA 1 [15]
14-23) Awards Total [15]

Section III - TRAINING AND EDUCATION DATA

24) Weapons qualification M 200605 [25] 27) APFT score 201 / P / 200810 [16]
26) Other resident training weeks 14 [70] 29) Self-development course hours 250 [50]
25) Post-secondary semester hours 60 HIGH SCHOOL DIPLOMA [60]

Section IV - VERIFICATION

a. I have verified this information. 29) I do [ ] I do not [ ] want to be considered for military education and promotion
b. IAW my enlistment/reenlistment contract, I have verified this information. I understand that if I decline a position within my selected region, I will be removed from the Enlisted Promotion List for the remainder of the life of that list. I understand that if I am on the Enlisted Promotion List, I may be scheduled for my required NCO Education by the state training office at the earliest possible date and will attend scheduled training. I further underst[and] if I fail to attend or complete scheduled NCOES training I may be referred to a reduction board per NGR 600-200 para 11-60. I also understand that [if] reduced in grade I will be moved into a position commensurate with my current grade. If a valid position is not available and I must be excess, I under[stand] that I will loose any bonus or incentives I may have effective the date of reassignment.

Soldier's signature and date

Section V - APPRAISAL

30) Board Member [ ] 31) Board Member [ ]
32) Board Member [ ] 33) Board Member [ ]
34) Board Member [ ]

Section VI - TOTAL SCORE AND VERIFICATION

a. Field 8	Time in grade	[10]	75 points maximum
b. Field 9	Time in service	[24]	75 points maximum
c. Fields 14-23	Awards	[15]	75 points maximum
d. Field 24	Weapons qualification	[25]	75 points maximum
e. Field 25	APFT	[16]	75 points maximum
f. Field 26	Other resident courses	[70]	75 points maximum
g. Field 27	Self-development courses	[50]	75 points maximum
h. Field 28	Post-secondary courses	[60]	75 points maximum
i. Fields 30-34	Promotion Board Score	[ ]	400 points maximum
	Total: Admin Points	[270]	825 points maximum

Printed as of 5/4/2009 8:23:04 PM

Recorder: Printed name, signature, and date Verifier: Printed name, signature, and date

NGB Form 4100-1-R-E / 1 Jan 02 Previous editions of this form are obsolete.

This is odd because this was not what I signed in Alabama, but it is what the 1-108th sent in from Alabama. Obviously, it had been thrown away and replaced with this. Why would they do that? Why would the administrative section of the 1-108th CAV forge a document? Even more telling was the standby form done by someone in Calhoun while I was in Afghanistan. That is not disturbing at all is it? I had filled out one of these while in Afghanistan. Was this done by the folks in Calhoun to cover their tracks? Or were they just doing it to do it? There was another mystery document changing my MOS even though I had been told many times that it had already been changed. I had even been given paperwork saying that the MOS had been changed.

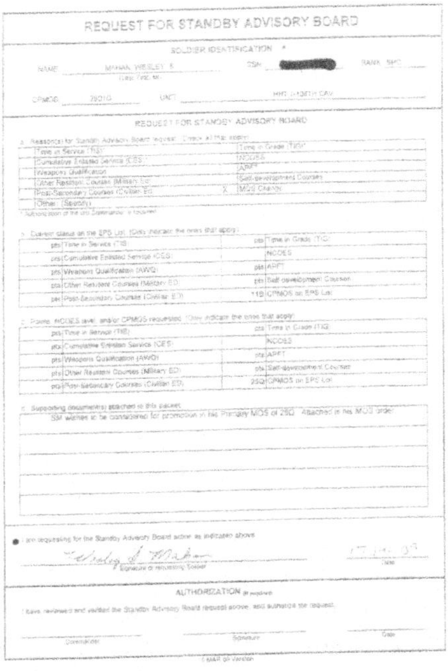

REQUEST FOR STANDBY ADVISORY BOARD

SOLDIER IDENTIFICATION *

NAME MAHAN, WESLEY S (Last, First, MI) SSN [redacted] RANK SPC

CPMOS 25Q10 UNIT HHC 1-108TH CAV

REQUEST FOR STANDBY ADVISORY BOARD

a. Reason(s) for Standby Advisory Board request. (Check all that apply)

	Time in Service (TIS)*		Time in Grade (TIG)*
	Cumulative Enlisted Service (CES)*		NCOES
	Weapons Qualification		APFT
	Other Resident Courses (Military Ed)		Self-development Courses
	Post-Secondary Courses (Civilian Ed)	X	MOS Change
	Other: (Specify)		

* Authorization of the Unit Commander is required

b. Current status on the EPS List. (Only indicate the ones that apply)

pts	Time in Service (TIS)	pts	Time in Grade (TIG)
pts	Cumulative Enlisted Service (CES)		NCOES
pts	Weapons Qualification (AWQ)	pts	APFT
pts	Other Resident Courses (Military ED)	pts	Self-development Courses
pts	Post-Secondary Courses (Civilian ED)	11B	CPMOS on EPS List

c. Points, NCOES level, and/or CPMOS requested. (Only indicate the ones that apply)

pts	Time in Service (TIS)	pts	Time in Grade (TIG)
pts	Cumulative Enlisted Service (CES)		NCOES
pts	Weapons Qualification (AWQ)	pts	APFT
pts	Other Resident Courses (Military ED)	pts	Self-development Courses
pts	Post-Secondary Courses (Civilian ED)	25Q	CPMOS on EPS List

d. Supporting document(s) attached to this packet.

SM wishes to be considered for promotion in his Primary MOS of 25Q. Attached is his MOS order.

I am requesting for the Standby Advisory Board action as indicated above.

Signature of requesting Soldier   Date

AUTHORIZATION (if required)

I have reviewed and verified the Standby Advisory Board request above, and authorize the request.

Commander   Signature   Date

When I transferred into the Georgia Army National Guard, I was sent to infantry school then signed up for pathfinder school with some potential class dates being given to me. However, I was retrained in Sat/Com. The reason for that was because I was hit by a drunk driver soon after graduating 11b school. A decision was made that rather than let me recover from the wreck i would be retrained as a 25q. To their credit the injuries were such that I never returned to what I was. Ever. The administrative section assured me everything was in order and that nothing adverse would happen to my records going forward. This was amid the ongoing pay problem I had for the first 13 months I was with the 1-108th. I was not exactly filled with warmth. The lack of warmth would prove to be a harbinger.

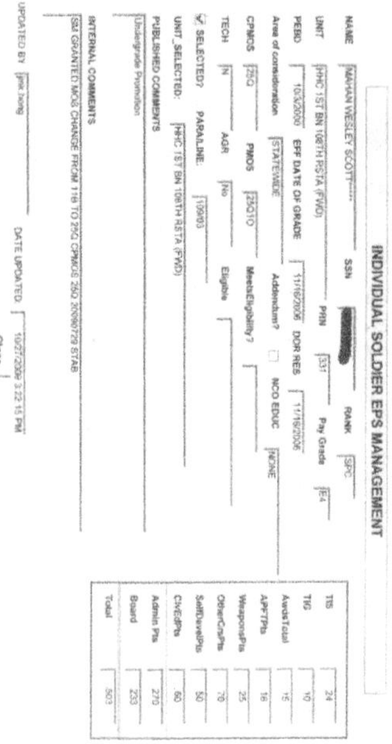

INDIVIDUAL SOLDIER EPS MANAGEMENT

NAME: MAHAN WESLEY SCOTT SSN: [redacted] RANK: SPC

UNIT: HHC 1ST BN 108TH RSTA (FWD) PRN: 331 Pay Grade: E4

PEBD: 10/3/2000 EFF DATE OF GRADE: 11/16/2006 DOR RES: 11/16/2006

Area of consideration: STATEWIDE Addendum? NCO EDUC: NONE

CPMOS: 25Q PMOS: 25Q10 MeetsEligibility?

TECH: N AGR: No Eligible

SELECTED? PARALINE: 109/03

UNIT_SELECTED: HHC 1ST BN 108TH RSTA (FWD)

PUBLISHED COMMENTS: Undergrade Promotion

INTERNAL COMMENTS: SM GRANTED MOS CHANGE FROM 11B TO 25Q CPMOS 25Q 20090729 STAB

UPDATED BY: [illegible] DATE UPDATED: 10/27/2009 3:22:15 PM

Close

TIS	24
TIG	10
AwdsTotal	15
APFTPts	16
WeaponsPts	25
OtherCrsPts	70
SelfDevelPts	50
CivEdPts	60
Admin Pts	270
Board	233
Total	503

So, when we were at Fort Mcclellan Alabama training to go to Afghanistan to end the war, that was what the leadership was saying, I filled out and signed an NGB form 4100-1-R-E for promotion. It had 25q on it. That seemed simple, and all the needed support paperwork had been turned in so there should not have been an issue. So, a few months later we are all in Afghanistan, the war did not end, and the promotion lists come out. Imagine my surprise when I was listed under 11b and not 25q! It was very upsetting. I would have been on the top of the list for the entire state on the 25q list while on the 11b list I was in the hundreds. The all knowing all caring administrative section of the 1-108th assured me it was a problem from Atlanta and not to worry about it. So, me not getting promoted due to someone else messing with my records is nothing to worry about? My leave was coming up that month and as I would be in Georgia, I decided to go to Atlanta to see what really got sent in. What I found there really changed everything. There was a problem, but it was not in Atlanta. The problem was sitting at a desk in Calhoun.

The form 4100 I turned in had not been sent in. What had been sent in was an unsigned form 4100 with 11b instead of 25q on it. Unsigned. This was not acceptable. I went to the IG office at a nearby FOB that was staffed by people from the Georgia National Guard. I showed them what I had and filed a report, told them getting promoted was the only thing I cared about at that time. We were in a war zone after all, and some things could be dealt with after the deployment.

After the deployment I found a standby form that looked to have been signed in a hurry that had me listed as a 25q, the standby form had been done after the NGB form 4100-1-R-E and after I complained. My signature had been forged on the standby sheet. This was disturbing. I had filled one of these out in Afghanistan before I went on leave so why did someone from Calhoun fill one out and forge my signature on it? Why was the paper that was filled out in Alabama not turned in in the first place? I thought there would be easy answers to those questions. Little did I know how wrong I was.

Most disturbing was that someone from Calhoun filled out a document about me and forged my signature on it while I was in Afghanistan, and this was seen as normal. That level of control over a person is disturbing to me. How does any form or contract that is presented to a service member matter if the 1-108th can rewrite anything and forge the signatures on it? That cannot be considered normal anywhere. How is that even legal or safe? Well, they won't say why but they will not stop either, so it is what it is.

The worst part of that was the reaction of the leadership within the 1-108th. The form I turned in in Alabama had been thrown away and replaced by someone in administration. Oh, I am sorry that is just a wild and unfounded accusation on my part. All the papers with forged information and signatures just appeared by magic. The replacement documents just appeared by magic as well according to the 1-108th. The first replacement had me listed in the wrong career field and was unsigned as well as undated. Then instead of admitting to it the 1-108th lies to me while we were in Afghanistan and after I solved the problem. There was no anger towards whoever forged the document and signature instead the anger was towards me, the victim, the person that reported the crime. They were mad at me for telling the IG that one of the full-time national guard members at the unit had thrown away the paperwork and forged it. It is a crime to report crimes in the 1-108th.

Nothing happens in a void. There was no hesitation in throwing away paperwork and replacing them with fakes. Fakes created by the administrative section of the 1-108th. After the fact, the finger pointing, and anger directed towards me was out of control but unintentionally tragic. Were the command and administrative sections of the 1-108th so used to people not reporting crime that they saw it as virtuous to attack a whistle blower? How out of control could they get?

There is an interesting postscript to that involving a travel voucher. Prior to going to Afghanistan, I had been sent to fort bliss in Texas for some additional job training. As such I filled out a DD form 1351-2 also known as a travel voucher when I returned to Calhoun. When the travel pay came in mine was lower than it should have been. This gave me cause for pause but with everything else going on it was not something I wanted to deal with at that time. However, when we got back from Afghanistan, I was in just the mood to see what went wrong with that voucher, I got a surprise. The one I had turned in had not been sent down to Atlanta. What were the odds of that happening more than once?

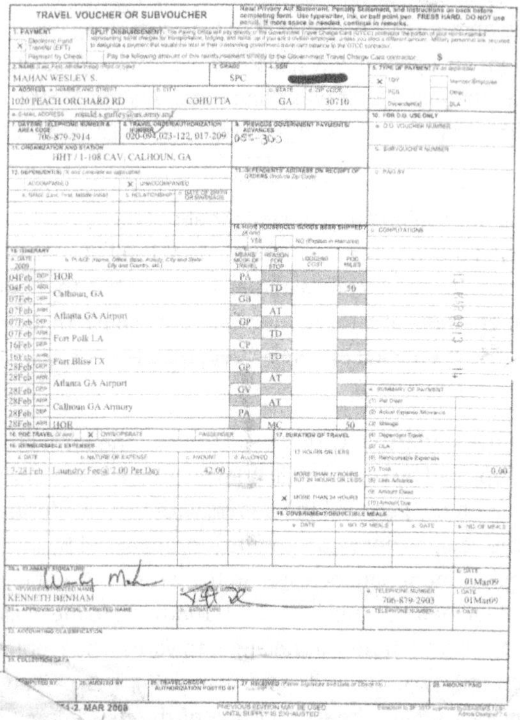

TRAVEL VOUCHER OR SUBVOUCHER

Read Privacy Act Statement, Penalty Statement, and Instructions on back before completing form. Use typewriter, ink, or ball point pen. PRESS HARD. DO NOT use pencil. If more space is needed, continue in remarks.

1. PAYMENT — X Electronic Fund Transfer (EFT)

2. NAME: MAHAN WESLEY S. 3. GRADE: SPC 4. SSN: [redacted]

5. TYPE OF PAYMENT: X TDY

6. ADDRESS: 1020 PEACH ORCHARD RD — COHUTTA — GA — 30710

7. DAYTIME TELEPHONE NUMBER & AREA CODE: 706-879-2914

8. TRAVEL ORDER/AUTHORIZATION NUMBER: 020-094, 023-122, 017-209

9. PREVIOUS GOVERNMENT PAYMENTS/ADVANCES: 055-300

11. ORGANIZATION AND STATION: HHT / 1-108 CAV, CALHOUN, GA

12. DEPENDENT(S): X UNACCOMPANIED

15. ITINERARY

a. DATE 2009		b. PLACE	c. MEANS/MODE OF TRAVEL	d. REASON FOR STOP	e. LODGING COST	f. POC MILES
04Feb	DEP	HOR	PA			
04Feb	ARR	Calhoun, GA		TD		50
07Feb	DEP		GA			
07Feb	ARR	Atlanta GA Airport		AT		
07Feb	DEP		GP			
07Feb	ARR	Fort Polk LA		TD		
16Feb	DEP		CP			
16Feb	ARR	Fort Bliss TX		TD		
28Feb	DEP		GP			
28Feb	ARR	Atlanta GA Airport		AT		
28Feb	DEP		GV			
28Feb	ARR	Calhoun GA Armory		AT		
28Feb	DEP		PA			
28Feb	ARR	HOR		MC		50

16. POC TRAVEL: X OWN/OPERATE

17. DURATION OF TRAVEL: X MORE THAN 24 HOURS

18. REIMBURSABLE EXPENSES

a. DATE	b. NATURE OF EXPENSE	c. AMOUNT	d. ALLOWED
7-28 Feb	Laundry Fee@ 2.00 Per Day	42.00	

(7) Total 0.00

20a. CLAIMANT SIGNATURE — b. DATE 01Mar09

c. REVIEWER'S PRINTED NAME: KENNETH BENHAM — e. TELEPHONE NUMBER 706-879-2903 — f. DATE 01Mar09

DD Form 1351-2, MAR 2008 — PREVIOUS EDITION MAY BE USED UNTIL SUPPLY IS EXHAUSTED

The voucher that had been sent into Atlanta was not the one I had turned in. Now I do not know who did that but my forged signature on the form looks like the clerk's signature that approved the voucher. What are the odds of that? I also found something unexpected, an additional NGB 4100-1-R-E. This one me listed as 25q, had been turned in while I was in Afghanistan and was also unsigned and undated. So, someone was trying to cover tracks even though they were still forging signatures during that period. That takes a certain amount of disrespect for the law and for the people whose signatures are being forged.

This is normal. The Georgia National Guard would be shown proof of a crime and choose to either ignore the crime or lie about the person reporting the crime. By the time all was said and done there were a lot more questions. How or why do multiple officers and investigators in the Georgia National Guard consistently look the other way over and over? Is it an unwritten rule in the Georgia National Guard that officers are allowed to commit and cover up crimes? Or was this just a case of several corrupt individuals just happening to be in positions of power and influence at the same time?

How hard is it to answer a few simple questions? When I say investigators, I mean members of the Georgia National Guard's Inspector General office and not criminal investigators. The Georgia National Guard has no CID or otherwise internal affairs office beyond the inspector general's office. In this case the Georgia National Guard's Inspector General office made it noticeably clear to me they would not investigate or prosecute anyone involved with these crimes. Why might one ask? Eventually the answer will come out even though anyone that spent time in the military already knows why. The Inspector General Office of the Georgia National Guard will not punish criminals of high rank.

So how much Integrity and Duty did it take to throw away those documents and replace them with fakes? Did the definitions suddenly change? There is also a glaring lack of respect in that as well. The administrative section thinks so little of service members that they have no problem forging documents and signatures. Not a good sign.

Chapter 2: Rules for thee

QUESTIONS!!!!

1. Why did randall simmons tell senator Chambliss(twice) that I had an affair with a female that he(simmons) made up? It's a crime to lie to a senator. Then add to that lying to the senator with the lie being that I committed a crime...

2. Why did randall simmons tell the senator forging signatures was not a crime and anyone could do it?

3. Why was my signature forged multiple times? (I don't have an exact count because the Georgia guard has no record of most of the freedom of information act requests, I sent in)

4. Why did he illegally punish me?

5. Is the Georgia National Guard accountable to anyone?

Those might seem simple but getting them answered one way or the other was a long and crazy journey. It included threats, lies and officers of the Georgia National Guard sticking their hands over their ears and screaming that they cannot hear me. Not literally of course as I was not present to see what they were doing. But through it all a few things remained true, the dishonesty of randall simmons who is a general in the army, the refusal of the army to break out of the British class system and my refusal to let them break my will. But it starts with the pay errors.

When I transferred to the Georgia National Guard, I thought nothing of it, I was in much better shape then and had been testing myself on the army pt scale for a few years, so it was the same to me. However, on my first pay stub I noticed something amiss. My pay entry date was October third, 2000, with no breaks in service, however the guard had that date as 6-27-2002! I was being underpaid. Now one might think this is an easy fix, just show all the proper paperwork to personnel and it gets corrected, and I get my back pay, right? Wrong! I turned in the papers, spoke to administration and was promised that it would get fixed. Was it? No.

When I went back to the unit, they were just shocked that it was not fixed! How could this be? I was told it would really be fixed this time and that promise became a lie. The administrative people just shrugged their shoulders, it was not their paycheck that was short after all. Anyway, they said it was only a little bit of money that it was off by. It was not their pay that was short. At this point I had seen enough, i went to representative Deal's office and filed a complaint. The Georgia National Guard responded with a letter. The letter went like this:

DEPARTMENT OF THE ARMY
GEORGIA ARMY NATIONAL GUARD
OGLETHORPE ARMORY
5019 HIGHWAY 42
ELLENWOOD, GEORGIA 30294-3438

August 14, 2007

Military Personnel Office
Georgia Army National Guard

SPC Wesley Mahan
P.O. Box 125
Cohutta, GA 30710

Dear SPC Mahan,

This letter is in response to your request for assistance through Representative Deal's office concerning your Pay Entry Basic Date (PEBD).

When you entered the Georgia Army National Guard your active duty Air Force time was not added to your Retirement Point History Statement. By updating this document, your PEBD will also be adjusted. I was able to use the scanned copy of your DD 214 from iPERMS to account for your active duty time and your USAF Reserve duty. I have enclosed a copy for your records to show you that your service began in 2000.

After going through all of your scanned documents and talking to the personnel at your unit I found orders regarding an Air Force mobilization. However, these documents do not establish active duty credit like a DD 214. If you can provide any paperwork that will account for active duty time while mobilized we can fix your status between 23 February 2005 and 15 November 2006.

If you should have any further questions or concerns, feel free to contact me at (678) 569-5304.

EDWIN A. LASTER
MAJ, AV, GAARNG
Deputy G1

Major Edwin A. Laster states that when I entered the 1-108th somebody forgot to put my active-duty time into the system. That sounds very professional however it is a lie. The recruiter that worked my transfer was very detail oriented and open about the process. He showed me everything and was very precise on my records. If Major Edwin A. Laster was telling the truth, then the error was even bigger than I thought. For what Major Edwin A. Laster said to be true that would mean that after the fact someone messed with my records. There is no other alternative. The paperwork the recruiter sent in correctly listed the transfer as seamless with no break in service. So, somebody either in the 1-108th or Atlanta went in and changed my records. For reasons unknown Major Laster does not mention this in his letter instead choosing the dishonesty route.

Why 6-27-2002 was chosen was never explained to me. The letter also states that documents regarding a mobilization were found at the unit, the only documents they had were two DD-214s that I gave them, this letter sets an unfortunately terrifying standard for the Georgia National Guard. That also made no sense because the recruiter sent the 214s in with my packet, he showed me before he sealed it and mailed it. Was the change magic?

The terrifying reality of the 1-108th is this: Truth does not matter, and we will lie to cover up crime. I was not activated between February 2005 and November 2006, that was not even part of the complaint. The complaint was that my entry date got moved from 2000 to 2002 thus making me be underpaid on every single check i got from them. There was never a claim of being on active duty between February of 2005 and November of 2006 but for an unknown reason Major Edwin A. Laster chose to say I had. This was either a mistake or a lie. Either Major Edwin A. Laster did not investigate the issue, or he intentionally lied to a service member and an elected official.

When it is shown that the 1-108th has made a mistake and they lie about the service member that reports the mistake why would anyone feel safe there? I was never told why Major Edwin A. Laster lied to the elected official about the situation. By his own statement, my entry date should have been 2007 not 2000. That made no sense and was obviously not true. Sadly, that would become a trend. The 1-108th and the Georgia National Guard would show to have no problem at all with telling lies about and to a service member.

Someone made a mistake with my paperwork with the result being that my pay was short. That is the polite way of saying it. Whoever that person was they must have had a lot of political power because no matter how many times i gave the personnel section of the 1-108th the correct paperwork the problem was never solved. Who was this person and why did they wield so much power? Of course, asking that question led to me being told I am they problem. What is the real problem? Are service members to expect never ending pay shortages? How can one person who is untraceable have so much sway over the records and pay of so many service members in the Georgia National Guard?

My pay was short and had been short since I transferred into the 1-108th. The chain of command was given the information and paperwork needed to fix the issue multiple times, but they refused to. I was forced to go to an elected official because there was no other way to get the pay fixed. What was the result? A letter written to me and an elected official which avoids talking about the problem and lies to the service member and the elected official! How hard is it to say your pay date was messed up and we fixed it? Instead, we get a rambling letter that talks about things that were not even part of the problem or complaint and attempts to subtly blame the victim. This will become a pattern. Not even fixing the mistake will also become a pattern! The Major said it was fixed. That was a lie as well!

He speaks of paperwork regarding an Air Force activation, but I really cannot grasp what he is getting at. There was no activation or active-duty time beyond early 2006. That was never even brought up by me because it did not exist. When I read this my first thought was that this guy did not actually do anything except write a letter with random and made-up points thrown in. This was a letter to an elected official as well as me! Respect? It is not there. Even when admitting that the Georgia National Guard somehow made a mistake the major still must throw a dig in. He implies it was my fault by bringing up an activation that never took place! Why?

Looking at pay stubs from November and December of 2007 does not help the matter. Remember that the pay error was fixed according to Major Edwin A Laster. Obviously, he lied. The letter was written stating that the problem was fixed and there was no mention of back pay. I go to drill and in a week and a half later get paid and the problem is not fixed. How is that possible? Really wanting to be nice here so I give them more time. Fast forward to right at a year of having been in the guard and it's still not fixed. But some unknown person in some unknown place in Atlanta is really working on getting it fixed. Really, they are. But were they?

During my time with the 1-108th I saw many people have problems just like mine. The problems were not all pay related as some were records related. That brought up a big red flag that I ignored. The Georgia National Guard had zero problem lying to a representative's office about fixing my pay. Honestly, I dropped the ball on my lack of reaction to that. Members of the Georgia National Guard lying to elected officials will become a recurring theme sadly. The only answer I have gotten about that is that they are the Georgia National Guard, and they can do what they want. It does not matter who the lie is told to or what they want to do, certain people in the Georgia National Guard are immune to the law, everyone else is just cannon fodder.

Alack of Integrity is obvious in Laster's letter. What does OCGA say about this. The Georgia National Guard does like to throw OCGA around after all.

O.C.G.A. § 38-2-511

GEORGIA CODE

*** Current Through the 2010 Regular Session ***
*** Annotations Current Through March 14, 2011 ***

TITLE 38. MILITARY, EMERGENCY MANAGEMENT, AND VETERANS AFFAIRS
CHAPTER 2. MILITARY AFFAIRS
ARTICLE 5. CODE OF MILITARY JUSTICE
PART 10. PUNITIVE PROVISIONS

O.C.G.A. § 38-2-511 (2011)

§ 38-2-511. Accessory after the fact

Any person subject to this article who, knowing that an offense punishable by this article has been committed, receives, comforts, or assists the offender in order to hinder or prevent his apprehension, trial, or punishment shall be punished as a court-martial may direct.

HISTORY: Ga. L. 1955, p. 10, § 60.74.

FOCUS™ Terms Search Within Original Results (1 - 1) Advanced... View Tutorial

View: Full | Custom 1 of 1 Book Browse Pages: 2

O.C.G.A. § 38-2-525 (Copy w/ Cite)

O.C.G.A. § 38-2-525

GEORGIA CODE

*** Current Through the 2010 Regular Session ***
*** Annotations Current Through March 14, 2011 ***

TITLE 38. MILITARY, EMERGENCY MANAGEMENT, AND VETERANS AFFAIRS
CHAPTER 2. MILITARY AFFAIRS
ARTICLE 5. CODE OF MILITARY JUSTICE
PART 10. PUNITIVE PROVISIONS

O.C.G.A. § 38-2-525 (2011)

§ 38-2-525. Failure to obey order or regulation

Any person subject to this article who:

(1) Violates or fails to obey any lawful general order or regulation;

(2) Having knowledge of any other lawful order issued by a member of the armed forces, which it is his duty to obey, fails to obey the same; or

(3) Is derelict in the performance of his duties

shall be punished as a court-martial may direct.

HISTORY: Ga. L. 1955, p. 10, § 60.88.

View: Full | Custom 1 of 1 Book Browse

O.C.G.A. § 38-2-525 (Copy w/ Cite) Pages: 2

Now as we see the things that have taken place in the 1-108th and Georgia Army National Guard so far are not legal. UCMJ applies to active-duty personnel or personal on active-duty orders, state orders and drill status fall under OCGA. I was sure that the Georgia National Guard would follow the law. I mean it would be book worthy if the Georgia National Guard intentionally broke the law.

Chapter 3: We wanted a long weekend

So now I am at Fort Gordon for 25Q training and going into the fourth week I have received NO pay. Now state of Georgia orders are processed differently than federal orders. Sometimes the pay processes differently and does not hit exactly at midnight or on any set day. So, I waited. Waited one week, then two, then three but received nothing. On Monday of the last week of class I expect pay. The pay does not get my account later that day or at all. It starts to look like I am not going to get paid at all. How is this possible? Could it be possible that the 1-108th totally forgot that I was at Fort Gordon?

On a Thursday I call the Calhoun armory and get no answer at any of the numbers. I called my platoon sergeant on his cell phone, and he said he would call somebody. After he makes some calls, he gets back in touch with me and lets me know that the armory is on a 4-day weekend, and they forgot to process my pay for the trip. What a shock. They forgot I was going to Fort Gordon. Forgot, forgetting, misplaced, eventually it seemed like the 1-108th used a list to find diverse ways to hold the 1-108th blameless. No matter what happens it is always just a big mistake that will never happen again. But it does! How does a command forget where someone is? I was given orders to be there, but those orders were never sent in to process the pay!

So, there was no way any pay would be in my account until Tuesday since if it were done Monday, it would take at least a day to process. That was all it took. I emailed state representative Tom Dickson, explained the situation, and added that my pay had been short for an entire year. Three hours later my phone starts to ring a lot, there was something that could be done after all. As a bonus the administrative sergeant that had to answer for that lied and said they knew nothing of my pay error. He forgot about the letter from Major Edwin A Laster. I know the sergeant lied because the Major that called me suddenly got a whole lot nicer when I told him my pay had been messed up since I got there.

I know there was no way the sergeant would lie to anyone calling from the state capital, right? Wrong, when asked about my pay being short for an entire year whoever answered the phone in Calhoun lied to whoever called from Tom Dickson's office, even though I had letters and paperwork showing that I was telling the truth! For shame! The 1-108th did not even blink they just lied to an elected official. Sadly, that would happen a few more times during my dealings with the 1-108th, they had no problem telling fibs to elected officials.

To put that another way the 1-108th said I never reported the pay errors and never brought in the paperwork needed to fix the pay errors. They acted like the letter to representative Deal's office had never been written, the amount of dishonesty that took was staggering. Yet another warning sign I did not heed. I had one person confront me at the Calhoun armory, when I pointed out that they pay error had been ongoing for an entire year this individual accused me of lying and walked off. I do not think he liked it when I pulled out copies of pay stubs and the letter from major Laster. I was prepared. It hurt his feelings to be caught in a lie by lower enlisted.

That illustrated the problem perfectly. He mumbled something about chain of command needing more time as he walked off. That would become a theme when dealing with the Georgia National Guard. Is honesty that hard and is pride that important? One year is not enough time to fix a pay error. What?! It was also OK for the 1-108th to lie to an elected official. If they can do that, what can they not do?

Even worse was the admission that the 1-108th had forgotten that I was at Fort Gordon! How do you forget where someone in your unit is? That takes a big amount of not giving a damn. Warnings, warnings everywhere and with the warnings and I did not heed them. So, the day before I left Fort Gordon, I got paid for being at Fort Gordon. I got some of the back pay but not all of it, that was too much to ask.

This whole incident was really a capsule of the 1-108th up to that point. They could not do what they were supposed to do, and it was never their fault. At this point I put in for a transfer as I had seen enough but my sense of honor nagged at me. On the one hand the unit I was in was obviously run by either incompetent or corrupt individuals and on the other hand they had spent a lot of money getting me ready to deploy, choices, right? After a sit down with two individuals I decided to stay.

At that point I already made up my mind that I would not be reenlisting after the deployment unless I transferred out of the 1-108th. If they could forget where someone was in the states what would they forget overseas? If something of that magnitude was forgotten overseas how would the 1-108th react? I had to ask. That question would be answered sadly. Integrity, Duty, Respect...all those things mean nothing? To put it bluntly a state employee at the Calhoun armory lied to whoever called from the state capital. Our state tax dollars at work.At no point while I was at Fort Gordon or after I got back to Calhoun did the phrase "we messed up" come from anyone in the 1-108th.

Chapter 4: It is Magic!

So, my pay was short for right at a year, the chain of command refused to fix it, an officer lied to a united states representative saying it was fixed and I am the bad guy? Then when I get back to Calhoun, I was yelled at for not given looking the chain of command enough time, because one year is not enough time? When I showed him the letter the discussion ended immediately. The "leader" just walked off, did not even apologize. That is only an appetizer.

It was a mirrored situation with the travel voucher from Fort Bliss. Although It was a normal voucher the signature on it was not mine and it did not have the per diem for being at Fort Bliss Texas on it. That resulted in me being paid less than I should have been, that was not the voucher I turned in. That one just vanished and was not sent it. I wasn't told that they lost it they just filled one out, forged my signature, and sent it in. The one I turned in just vanished.

How do papers that I turned in for pay and promotion get thrown away or lost and replaced with a paper with incorrect information on it and i am never told? So, when the lists come out while we are in Afghanistan and i am listed under the wrong career field and it is no big deal. I was not the only person that this happened to nor be I the only person who had trouble with travel vouchers and pay but I was the only person who let the 1-108th know that they were not doing their jobs. The usual line they gave was "that is just the way the military was" which when they were dealing with people who had zero prior military service experience could kind of work. But there were some of us who had served in other areas of the military and could see where the problem was.

Either the administration section of the 1-108th was dishonest or the administration section of the Georgia National Guard in Atlanta was incompetent and dishonest. Unfortunately, it was a non-secret that the 1-108th was lazy and dishonest. It was their culture. When people who had served in active duty or different units spoke up, they were told that they were wrong. The 1-108th had to maintain the illusion that crime was normal. Anyone that brought up the fact that the 1-108th was a borderline criminal organization was attacked or called stupid. But what would they do when someone they attacked just welcomed it?

The culture of the 1-108th or the Georgia National Guard was...interesting? It was obvious they were used to dealing with people right out of high school or with little or no prior military service. At one point a jag lawyer responded to the proof of an officer telling lies about me to an elected official by threatening me with criminal charges, things you should never email part infinity. So, what was the upshot of all that? The administrative section of the 1-108th loses some but not all the paperwork turned into them and instead of telling the service members just forge documents and signatures to fill in the gaps? These cases of travel vouchers and other documents being forged were not unique to me as others also had that happen to them. Is it only the 1-108th that forgers documents and signatures on a regular basis or does the entire Georgia National Guard do so? That is a simple question. That question also upset a few people.

Like any good magician the 1-108th became terribly upset over their magic tricks being exposed. In fact, one might say some of them became a little unhinged. The anger however was not directed at the individuals performing the magic tricks but at the victims of the tricks! What backwards criminally slanted mentality is this? This is a complete failure of leadership. When a service member reports multiple crimes, the correct response is never to threaten the person reporting the crimes. But that was the way the 1-108th operated. Luckily, they kind enough to give a good example.

The threat email was after Afghanistan, first sergeant brian green was always back and forth on his moods and on that day, me asking for a transfer out of there must have set him off. I never saw or heard from him again after receiving this email, he was not at any of the drills and no reason for his absence was given. I thought the 1-108th was big on attendance. Anyway, I would have loved for him to say those things to my face, but he never did. I do not know what kind of culture he was used to but when a person says those kinds of things via an email then avoids the person, they were threatening it makes me laugh. It also comes off as cowardly. I guess brian green could only threaten people via email. He had shown that pattern before. He always showed bluster if he had protection or was able to walk away quickly after running his mouth. Let us go over the kind email brian green sent out about me.

Page 2 of 4

A transfer is a request not a right, it can be approved or rejected for any reasonn or no reason.

SFC Roney McStotts
S-6 NCOIC
1-108TH CAV SQDN

Re: Fwd: Re: my forged signature.eml

Subject: Re: Fwd: Re: my forged signature
From: "Green, Brian K 1SG NG NG NGB" <brian.k.green@us.army.mil>
Date: Thu, 27 May 2010 08:32:21 -0400
To: "Mcstotts, Roney L SFC NG NG NGB" <roney.mcstotts@us.army.mil>, "Benham, Kenneth" <kenneth.benham@ga.ngb.army.mil>, "Enriquez, Carlos C CPT NG NG NGB" <carlos.enriquez@us.army.mil>, "Recker, Joseph E USA CSM USA 1-108 CAV CSM" <joseph.e.recker@afghan.swa.army.mil>

Gentlemen,

We all know that SGT Mahan is a Mal-Content, Whinny little bitch. He rarely remembers reality as it actually occurred. He has a long standing history of dreaming up a situation that would be most advantageous to him. I'm sick of this individual. He no longer deserves to be an NCO in my Army! I will not be in a position to affect this until early July. At that time I intend to spend all my time and energy getting SGT Mahan reduced, then chaptered out of the Army. Transferring this individual to another unit simply because we are tired of dealing with him is the wrong answer. I can be reached on my cell phone at 706-346-9532. If SGT Mahan needs me to restate this to him personally, please forward my number to him! I don't care who he has talked to. I don't care if he has a complaint to submit after I talk to him. I don't intend to hurt his feelings, I intend to end his military career.

1SG Green

----- Original Message -----
From: "Mcstotts, Roney L SFC NG NG NGB" <roney.mcstotts@us.army.mil>
Date: Tuesday, May 18, 2010 5:06
Subject: Fwd: Re: my forged signature
To: "Green, Brian" <brian.k.green@us.army.mil>

> We would like to send Mahan on his way but he has to talk to the
> Col first.. I have offered to drive him to the Col.. The paper
> work is ready just waiting for the signature. Is there anything
> you can do to expite the process.
>
> SFC Roney McStotts
> S-6 NCOIC
> 1-108TH CAV SQDN
> ----- Original Message -----
> From "Mahan, Wesley E SGT NG NG NGB" <

All I wanted to know was why my signature had been forged a few times and to leave, they responded by name-calling me denying the truth and giving out threat

https://webmail2.us.army.mil/iwc_static/layout/shell.html?lang=en-us&15.01_143243 11/10/2.

A "Mal-content", perhaps I was but I was furious at nothing being done. They had threatened to court martial me over a woman they made up, my signature had been forged a few times and the chain of command's reaction was to threaten me to say it could be worse. Could be worse? A definition of worse beyond document and signature forgery plus false charges is not something I want to see. Their hand had been shown there was nothing left they could do except kill me. All they wrote or said now came out to defending criminals, justifying fraud, making threats or being a criminal. Nothing they could say mattered anymore and I had zero problem saying that. I guess that made me a malcontent. This also showed their ignorance of my medical records, which were all lost.

Given that the letters written to the senator along with the email in which randall simmons admits to lying and attempts a cover up exist this is not a dream. On a side note, denial is a Freudian defense mechanism, I mean it's better than admitting the unit you oversee has criminals in it. What else do you call people who forge documents and signatures in order to control people? I can think of a few things, but they are not polite, and I am a polite person. Sadly, getting underpaid for the first year I was there was not made up, my signature being forged a few times was also not made up, I understand it might someone feel better to attack a person reporting crimes, but the crimes still took place. With the only course of action, the 1-108th would take was attacking me, the whistle blower, what did that say about them?

The intellectual dishonesty it takes to insult and threaten a whistle blower when you know the whistle blower is telling the truth is staggering. In fact, calling the truth a lie when one knows it's true is not what a leader in the military should do, it's what criminals do when they are saying everything, they have done is not their fault. If brian green had just done his job and not gotten mad because one of his friends was overseeing the forgery, I would respect him and I would have cared what he was saying about me. But integrity is a core value only when it is easy apparently because brian takes the side of the criminals here and promises to punish the person blowing the whistle on the crimes. That is not what a leader in the military should do.

I "no longer deserve to be an NCO in the army". That line might betray deeper problems. Does the army have highest suicide, divorce, and domestic violence rates in the military? Unreported or hidden it's all the same. Ask Fort Hood. Could service members having to deal with hostile and dishonest leadership have anything to do with those rates? He will "intend to spend all my time and energy getting SGT Mahan reduced, then chaptered out of the army...". That could be what is defined as winning game 1 then losing four games in a row and the series. Brian green and some others honestly, I thought I would suddenly forget about all this when I was out. How many other people who were victims of their crimes did they do that to?

My problem with that is that it became obvious very quickly that the 1-108th was not the army. The sad joke of his little rant was that my military career ended in august of 2004. If the 1-108th had not lost all my pre-1-108th medical records they would have known that i was just a ghost in the aftermath. He wants to end my career it is nice to have goals in life. Now is that ethical or legal? At the end of the day, he is saying all those things about a service member who used the chain of command to report crimes.

How ethical is it for someone in a position of leadership in the military to state they want to punish someone because that person reported crime to the chain of command? Do the classes and military education the Georgia Army National Guard sends people to tell them to do that? Would a leader not want to stop forgery and fraud not threaten a person reporting it? The legality of that comes into play as well. Is it legal for a leader in the military to threaten someone who is using the chain of command to report multiple crimes? Or does that violate the whistle blower protection act? Of course, as I was soon to find out the law does not apply to the 1-108th.

He sent that email to carlos the squadron commander, kenneth an administrative nco, joey the sergeant major and rony my platoon sergeant. Nobody had a problem with it what strong leadership with an illustration of great integrity that was! His misspelled bitch insult was so horrible i put it on a shirt. But being serious for a moment that email reveals a lot about his leadership style. A service member in his unit uses the chain of command to report crime and then requests a transfer and Brian Green's reaction is to threaten that service member. That is not exactly legal nor is it a unit i wish to be in.

While authoring that email, he knew that multiple pay errors had taken place, not all of them were corrected, my signature had been forged a few times with one of those times costing me around 2000 dollars and I had been told to admit to having an affair with a woman that did not exist, and not in a fantasy way. Those do upset a person. So, if I read his email right because I did not just roll over and let things happen to me, I am the crybaby? Apparently, I was making it all up to. So according to a leader in the 1-108th reporting crime makes me non-military material? Fort Hood would love him.

Is that what he does? Is that what he wants troops he is responsible for to do since that would make his job easier? Never found out because he never came to drill again while I was still in. Yes, that is sadly correct, after sending out that brave email about me he avoided me. By avoided i mean that he never went to a drill while i was still in, never contacted me again and would not pick up the phone. That was a little disappointing.

The email bothered me but not in the way he wanted. I reported some crimes, the result was a threat and an implication that they would make me look bad. Brian states that he doesn't care who I talked to because he wants to get me and that is the worst part. He knew or should have known that after being threatened by the 1-108th chain of command i went to the inspector general and a senator's office. So, what is Mr. Green saying?

He is stating that he wants to retaliate against an individual who blew the whistle on crimes. He then states that he does not care who the individual has spoken to. At the end of the day brian green would rather punish the person reporting the crime instead of punishing the criminal. That is not an organization I want to be in. This does not follow the core values of the Army at all. How could anyone in the United States army, even the Georgia National Guard variant, think it is ok to attack a whistle blower and to protect criminals? I wish so badly he had threatened me to my face, I know his type from when I worked bar security. It would have been worth it. Also, nobody ever offered to "drive me to the col", I was told I had to "talk to the captain" except he never had time to talk to me. This was a little odd because things had gotten out of control, and I wondered if they thought I would just say nothing in front of Randall.

Now let us say that i did not suffer a neck and shoulder injury in 2004. His threat and what the 1-108th eventually attempts to do could have had horrible consequences on me. But thankfully they threw away all my prior service medical records. No, I am wrong, they were not thrown away they were "lost", "shredded" or "burnt". One of those three things. The medical record went from my hand to the administrative NCO's hand. I never saw them again. I went to the Georgia National Guard IG with that email. They refused to do anything because the ranks of the criminals were too high. Fort Hood. Also, brian green tipped his hand. Now that I knew he had no respect for the law, and he wanted to run me off I had the advantage not the 1-108th.

Now these things were not unique to me, I saw several members of the 1-108th have problems just like mine while I was there. Everyone was told a variation of the same theme, whatever took place was not the 1-108th's fault but "someone in Atlanta". There were always the folks who complained but as far as i could tell the 1-108th had never dealt with someone who had served active duty prior to them and had no problem pointing out the criminal activity they allowed to happen. It honestly seemed like the 1-108th did not know how to react how to act in a legal manner. They only knew how to lie to people then follow up the lies with threats. When it got past that they got stuck in a pattern of lies, threats and denials. Why?

Unintentionally brian green also stated the 1-108ths motto: If we say something did not happen it did not happen. Sadly, the problems were not fake. When we were in Afghanistan and (at least) five people out of 30ish had their promotion paperwork messed up, was that fake? When travel vouchers paid out short constantly, for certain people, was that fake? Obviously not but anyone that mentions the crimes is attacked and called mean horrible names or at least they are supposed to be mean and horrible. I thought they were silly.

To lie about forgery and fraud when the people complaining have the original papers is stupid. That is a problem. An individual or military unit's refusal to prosecute members who forge documents and signatures does not bode well for anyone in that organization. Brian stated that himself when he said that I was making it all up. Of all the options at his disposal he chooses to attack the whistle blower, not good leadership form the Georgia Army National Guard.

One last thing about the email, which was the last communication I ever got from brian green. Obviously, I did not make anything up about forgery, fraud, or my pre-1-108th medical records all vanishing. His reaction was one of the reactions the 1-108th had with the other being denial. Now it made brian green feel better to insult and threaten me but that could not change two harsh truths. The first is that forgery and fraud did happen there and judging by the reactions i was getting they were considered "normal." The second was that if my transfer was not approved, I would be out within a few months, so any threats had a window to fit into.

A window that was getting smaller. All had to do was keep winding them up and collecting the replies, eventually they would be good for something. They would either transfer me or I would ETS. It seemed like the leadership of the 1-108th honestly expected me to just roll over and die for them. But when I did not write a false statement for them in Afghanistan that was a hint they did not see, we were not in Afghanistan anymore. We were back in Georgia on drill status. There is a little bit of a difference.

So how well did brian green use the core values? Did he treasure integrity? If that is defined as insulting and threatening a whistle blower then yes. Did he show Honor or Respect? Well, he did honor his commitment to protecting criminals and he respected the criminals right to have the victim removed so there is that. Personal Courage, how brave was he to threaten me, use his rank to sandbag me then never have the courage to say anything to my face or come to drill when I was there?

Although I never saw him again, I did hear from brian green one last time. He emailed me and told me that by asking the questions I was about my transfer request being sat on and the crimes that took place at the 1-108th I had violated the UCMJ. The UCMJ does not apply to service members on drill status or state orders. I was on drill status. I laughed when I read them.

After the helicopter incident in Afghanistan, I do not know why he would have sent me an email like that. What did he think I was going to do? Cry? Beg for mercy? Or print out the emails and drive down to Atlanta to see a jag lawyer. The lawyer really did not want to do anything, but he did chuckle and said he would call the 1-108th. I never heard from or saw brian green again. Such bravery.

Chapter 5: But I said to run into a minefield

A few quick examples from "brian green's army," they are funny in a way. While we were at Camp Shelby for training, we spent some time living on a mockup FOB to simulate being in a combat zone. One night there was a simulated bombing attack and brian green had previously said that in such an event we were to stop what we were doing and immediately go to a bomb shelter. No matter what. Even if you were in hard capped building 100 yards or more away from the shelter. I wonder if he expected casualties and if deaths were acceptable if people left a safe building to run to the approved shelters? Just like the power point slides in the meeting said they should. After all the shelters were there to protect people from getting hurt even if people had to leave a building that offered more protection, then the shelter. Where does this logic come from?

During the simulated attack several of us were in hardened buildings and logically we chose to stay in the buildings. Why would a person leave a building during an attack and run up to 100 yards to get to a shelter? Bombs would be falling. Now despite that logic we were greeted with anger after the attack was over. Brian Green was ranting about people not going to the shelters and not taking anything seriously as if he really expected people to leave a hard building and run 100 yards through a bombing attack to a shelter that offers less cover then the building! I made a point to say that and got an attempt at the evil eye. Yes, an attempt. He had bluster but was really a coward underneath it all.

The next week we are still at Camp Shelby and the entire unit is given some off time with one caveat delivered by brian green himself: Do not be late to formation in the morning, being late will have dreadful things happen. He seemed profoundly serious even implying that extra duty or loss of pay might be involved if one was late. Well, that morning rolled around and shockingly there were a few people that are not even awake yet when the formation assembled, this could be bad! What would the consequences be?

The individuals that overslept were in a platoon led by one of brian green's best pals. Now this presented a test of the professionalism of the 1-108th, would brian green follow through on his threat? Or would he play the buddy system card? Sadly, the answer was no. They were pulled into a meeting between brian green, their platoon sergeant and after the meeting they were laughing about getting away with it. What wonderful leadership. Things like that have a negative impact on unit morale and cohesion. If someone was worried about those things.

One night while we were living at a post near Kabul there was a rocket attack on the facility. By attack I mean one rocket that landed in an athletic field on the Afghan part of the compound. Many of the buildings there were hard buildings, brick foundations, brick walls you get the point. Most of us had already decided that we would not leave a hard building to run across a base during a bombing attack, it was not smart. Also, the attack was around 3 am and was only one rocket. Most people were asleep at that time.

Anyway, the next day word starts going around the base that leadership was upset because people did not leave hardened buildings and run across a base during a live rocket attack. The fact that the shelters had less shelter than the buildings was lost on them. There was talk of having a drill, in a combat zone in a base surrounded by mountains mind you, to make sure everyone would go to the shelters they were assigned to. Common sense beat the boardroom on that one and the idea of the drill got laughed away. Now his name was never attached to it, but it sounded like something he would say. The idea for the drill came from the leadership of the 1-108th. "Leadership" at least as it is defined by the 1-108th would rather people needlessly put in danger because of what? A policy? Is someone not feeling good about what they see in the mirror?

When we were at Fort Mcclellan Alabama getting ready to go to Afghanistan it was decided that to help morale we would have a fighting tournament. Fighting being a generic term we were supposed to use the army combative styles on each other. A bracket was written up with the lower enlisted on it. Middle and upper enlisted did not participate. Also, officers did not participate but they were all going to watch. We were going to fight for their entertainment. None of them saw the train wreck coming.

By that time clicks had formed with the unit, and it did not take long for a few of us to look at the bracket and realize one thing. This was going to present us with the opportunity to hurt some of the ass kissers in the unit and not get in trouble. Some agreements were made, the deal being that some of us would throw our matches to get a few of the ass kissers into the later rounds. They would get beat up but win the first fight and even the second, but they would lose the third and be out after taking three beatings. Sadly, the tournament was stopped after the second day. It might have been obvious.

When I was at Torkham while I was being accused of having an affair with a married woman that did not exist brian green arrived to do my performance evaluation. It was mediocre and mentioned the non-existent affair on it! I was expected to sign something with lies on it! That is just the way of the 1-108th I guess. I did not sign it with my name I wrote bullshit and left, I sometimes wonder if he even noticed or if he even cared. I would lean toward him not caring. Such a great leader in the Georgia Army National Guard this brian green is!

Previously, at Camp Shelby the accident that got me retrained came up in the physical one takes before you deploy. They wanted to get my left leg looked at just to make sure it was ok; this was only a formality as I had already been told I was going to be cleared to deploy. However, this did mean I had to break away from unit activities and go to an appointment. Brian Green reacted to this by telling me I was lazy and that he was going to make sure I did not deploy. He said that to my face. I replied by saying that i was not lazy and he responded by saying I was and then walked off. He walked off extremely fast at a jog. So, if I was hearing him right me being hit by a drunk driver makes me lazy? The extra medical appointment was to get me cleared to deploy. What a leader.

I made a few phone calls and told my platoon sergeant that if I was sent home from Camp Shelby there was another guard unit that had indicated that they would pick me up and I would deploy with them. The very next day brian green apologized for calling me lazy and said he would not move me out of the 1-108th's deployment team. Later, down the line it would become obvious that he never got over having to say those words. So, I went to the appointment got cleared and deployed with the 1-108th. What could possibly go wrong? In a way it seemed that green had wanted me to beg him to let me deploy. Why?

While we were at Fort Mccelean the rank-and-file troops were confined to the fort and range areas. I understand that for training reasons and because of where we were deploying to why this was done. People need to get used to living austere. But there was one little thing that just stood out. A Lot of the leadership was never around during evening chow hours, they just were not there. It turned out that they were using personal vehicles to go out into the local area on some nights. Rank has its privileges blah blah blah, but privilege does not mean abuse. It shows what leadership really thinks of people when the regular members are eating meals ready to eat or chow hall food and the leadership are dining at steakhouses.

Brian Green also showed his leadership skills at an airfield near the Afghanistan/Pakistan border. We were there to pick things up and other people had ancillary things to accomplish, I was one of the other people. I had to go to the com sec area of the base for some items, due to the nature of the area my phone was left outside on a desk. When I return there are no missed calls showing and the people there say the phone did not ring. I return, early, to the motor pit and all our vehicles are gone. Brian Green oversaw that convoy.

The military police drive up to me and tell me that the convoy left me, this is interesting. The military police also tell me that a few people asked about me, but the convoy commander(green) said they were leaving anyway. I call back to the FOB and they tell me the convoy is back, and nobody knows how I am supposed to get back. One person said he was “working on it”. Well, I traveled a lot during that deployment, I volunteered for everything I could to make time go by and not to be a coward. A side effect of that was that I knew a lot of people in charge of where troops sleep and how they travel, I kept that on the down low for my own reasons. There was a helicopter pad at a compound near the one that I lived in.

I went to flight ops and just stayed honest. I had no helmet, no body protection, those had been left in the convoy. After laughing at first, they realized I was serious and decided on a solution. There was one last flight for that day that had a few stops but not the area where I lived but in their kindness that post was handwritten onto the manifest and made legit. A few hours of flying later I call the section tell them I am at the pad and need a ride, after the initial disbelief I stated that if someone didn't come get me, I would walk.

There was only an Afghan police area and a small village between the two FOBs. I'd walked it before but that was with others as we were not supposed to travel between the FOBs alone. Strangely one person drove up on an electric cart to pick me up. ONE PERSON all alone. I was told brian green was furious. I was not supposed to find a way back I was supposed to be afraid and ask the 1-108th to save me. I was not supposed to fly back on a Candian helicopter.

I was supposed to stay there overnight to be taught a lesson. What lesson? This was after I had blown the whistle on the personnel section. I did not care what they thought because they had made it obvious, they did not care about me. Brian only spoke to me one more time after that day. His pride was wounded, and he could not understand how I got a flight back. He had no idea that most of the people at flight ops were from Canada and like me loved the Red Green show. Looking back on that now I think he also could not comprehend real teamwork and helping someone else out because it is the right thing to do. He could only see his need to hurt someone that reported crime.

Now the military police told me Brian green just told everyone to leave, when I get back to post I am told they tried to call me and failed. My phone showed no calls. Going beyond that brian green knew where I was at, they received no calls from anyone on that convoy. So, it is either believe the military police or believe a senior NCO that was rumored to take pills to calm his moods down, choices, choices. That was just a rumor that he took mood pills, he did show all the signs of it. He was very moody though and the moods matched the ups and downs of substance abuser. Which also match, usually, the pattern of someone addicted to mood pills. But what would I know? I am just a "whinny little bitch".

Speaking of flying my attention to detail ended an incident the 1-108th started mere minutes after it was brought up. I was approached by an officer who stated that I had lost some parts I had picked up from a FOB the previous week. I replied that I had not. The officer said that my information had been used to get the parts and as the FOB was a mile down the road I must have been there. The officer told me to find the parts. I went back to my bunk and looked in my journal. I then went to the sergeant directly under that officer and reminded him that on that day that my information had been used to get those parts I had been flying in another part of the country in a helicopter. The matter vanished immediately.

Of course, that behavior was normal for he 1-108th. One time while in a convoy going from Camp Phoenix to Camp Blackhorse a door on our humvee swung open and supplies started spilling out onto the road. Our vehicle called for a halt and a cloverleaf and the lead vehicle, which was staffed by one of brian green's best friends said why and everyone just kept on driving while we were stopped fixing a situation. Watching those hummers drive off and leave us i could only think of what would happen if we got swarmed. Thankfully, we gathered what we could and left the area before we could be attacked. We even caught up to the rest of the convoy who thought we had just slowed down despite a halt being called for. When we got back to Blackhorse, they just asked why our vehicle called for a combat halt. Um...because we might have needed help? This was reality not a drill weekend where people could just kiss ass and hide. Such Integrity and Personal Courage.

Another time while we were at Blackhorse there was a late afternoon call for a convoy to go to BAF and pick up an active-duty full bird. As the time was creeping into early evening it was reasonable to request night vision goggles for the gunners. This request was denied because the 1-108th supply NCO was worried something might happen to the goggles. The goggles were more important than any lives on the convoy apparently.

While we were at BAF the sun went down. Afghanistan being what it was the route we were on had no lights unless driving by a compound or "farm". I was in a gunner hatch. There was a point when while looking into a field I saw flashes of light. My immediate thought was muzzle flash. I locked in my aim and told the rest of the vehicle what I saw, the vehicle commander told me to do what I needed to do. A few seconds later I realized I was looking at a campfire and some nomads. Thankfully I didn't shoot at them, but some night vison goggles sure would have been nice.

Chapter 6: It is not a trap if you know it is a trap

That email from brian green really set the stage for everything that followed. For the first year I was with the 1-108th every check I got was short, did that benefit me? Even though the administration section had been given the paperwork needed to fix it multiple times they never did. Yet when called out on it they lied to representative Deal's office about it. In my mind I saw the 1-108th as either ran by idiot or ran by criminals, which one was it? While we were in Afghanistan, I let it be known I would put in for a transfer or just get out, I was done with them. The reaction was hostile and reminded me of a mafia movie. Also, this email did not scare me at all, since they wanted to play games, I was game for it. This is going to be fun.

The email indicates the major problem of the 1-108th and unfortunately in some cases the military in general. When they are faced with someone who points out their illegal actions, they attack the person instead of not performing the illegal actions anymore. Over time this pattern creates a horrible environment. The part about ending my career was tragically hilarious. When I transferred to the 1-108th they lost all my medical records from my military time, in those records was a little piece of information that would have gotten me medically put out of the military. I still wanted to transfer and was pissed about the lack of integrity in the 1-108th but you got to laugh.

It should have ended when the injury occurred in 2004 but after a lot of physical therapy and a few rounds on a traction machine I was given the choice to stay in and I did. After I got out, I went to the VA, they looked me over and I was immediately put on disability and told I could never go back into the military again. I would never be able to pass the physical due to the injury. Despite his harsh words I never saw brian green at drill or otherwise again. Now why would someone in a position of authority say all that about me but not to me? He seemed eager to talk about me but for some reason could never say anything to my face. I wonder why? Did he lack personal courage?

In fact, that email really upset me but probably not in the way he intended. I already knew that people in the military like to abuse their positions, it's an early lesson. But if you're going to lay out some threats and insults like that at least have the moral courage to come to a drill and say it to my face. Also, throughout my time in the military I found that people that called the military "mine" were mentally unbalanced. The military owned us not the other way around. Having seen brian green's leadership skills I wanted nothing to do with anything he was involved with. Oh, I wanted brian green to be at drill so much...but he never was after sending that. I was looking forward to seeing him. But sadly, I never would again, he just couldn't get to any drills while i was still in. I thought that was a big thing for the 1-108th, always have to be at a drill even if it's not a drill. Or maybe they just apply rules to certain people and then even to certain degrees.

LETTER OF INSTRUCTION – UNEXCUSED ABSENCE

TO: MAHAN WESLEY SCOTT 1020 PEACH ORCHARD ROAD COHUTTA, GA. 30710	FROM: COMMANDER HHT 300 NORTH RIVER ST. CALHOUN, GA. 30701

PART I. RECORD OF UNSATISFACTORY PARTICIPATION

This is to notify you of reason(s) you were recorded as unsatisfactory participation:

a. (X) Unauthorized Absence from the scheduled unit training assembly (UTA) or multiple unit training assembly (MUTA) period(s), 25 – 27 June, 2010 1☐ 2☐ 3☐ 4■ 5☐ 6☐

b. ( ) Charge with unexcused absence on ____________________ due to:
( ) Improper Military Appearance
( ) Unsatisfactory Performance of Assigned Duties

PART II. CONTACT ATTEMPTS

Record of contact attempts with absent soldier

DATE MM/DD	TIME	PHONE NUMBER OF SOLDIER	PHONE NUMBER OF NEXT OF KIN OR EMPLOYER	REMARKS	FIRST LINE LEADER NAME AND SIGNATURE
25 June 2010	0800	706-271-7125		See below	SFC MCSTOTTS
26 June 2010	0800	706-271-7125		See below	SFC MCSTOTTS
27 June 2010	0800	706-271-7125		See below	SFC MCSTOTTS

Results of phone conversation: SM stated he is unhappy with the Unit and does not feel he should drill.

PART III. COUNSELING

All counseling will be conducted using DA Form 4856 dated June 1999, Developmental Counseling Form. All counseling will be conducted in accordance with FM 22-100 and FM 22-101. A copy of DA Form 4856 will be attached to this form and mailed to the soldier if the soldier is not present for counseling.

The counseling must state the number of attempts to contact the soldier by phone to home of record or place of employment. If failure to contact the soldier at these locations results, further assistance will be accomplished by either contacting the Next of Kin or the local Law Enforcement agency.

PART IV. INSTRUCTIONS

1. Under the provisions of AR 135-91, you are required to attend all scheduled unit training assemblies and annual training periods. In addition, you are required to participate in a satisfactory manner with regard to proper military appearance and performance of assigned duties.

2. Unless absences indicated in paragraph 1 are excused, you will have accrued _4 unexcused absences within a one-year period. The one year period begins on the date you incur your first unexcused absence.

3. Absences from training assemblies may be excused only for reasons of sickness, injury, emergency, or other circumstances beyond your control. If your absence was for one of these reasons, you should furnish this unit an appropriate affidavit by the doctor, medical officer, or other person(s) having specific knowledge of emergency or circumstances, requesting that your absence be excused. Your absence cannot be excused unless your request and affidavit or certificates are received within 14 days of the date you receive this letter.

4. You will be notified in writing within 10 days after receipt of your request as to whether the absence has been excused.

5. If you have family responsibilities that are causing a hardship, or if your civilian job is of critical importance to the national or community health, safety, or interest, you should contact me so I can advise and assist you in the proper procedures to resolve these problems.

6. As you are aware, if you accumulate nine (9) unexcused absences within a one year period, you become an unsatisfactory participant and may be transferred to the Individual Ready Reserve (IRR) per AR 140-10 or NGR 600-200 for the balance of your obligation. Eligibility for an incentive prescribed by AR 135-7 terminates when a soldier becomes an unsatisfactory participant.

7. I hope that as a result of this letter, you will take immediate steps to improve your attendance.

8. The next scheduled training assembly for this unit is 0800 hours on 17 July, 2010.

CARLOS C. ENRIQUEZ
CPT, AR, GaARNG
Company Commander

Receipt of the Original Hereof is Acknowledged. Soldier Unavailable To Sign
SGT WESLEY MAHAN

PART V. Commander's Statement

SM is AWOL and refuses to report for drill, recommend reduction in rank/grade to SPC/E4.

GANG Form 0001-R dtd 1 October 2002
Previous editions of this form will not be used and are obsolete

The letter, it did not matter that the phone number was not mine and that nobody contacted me or my "next of kin" on those days, brian green's threat had to be made reality.

When I was speaking with carlos enriquez about this I got the impression he got turned on by his own power. I called him after receiving the letter (yes, we had each other's numbers) from him and asked what was going on because...the number he had listed as mine was not mine and nobody from the 1-108th called me on the days that captain enriquez wrote that I had been called. So, he lies in the letter and says he called me three times then lies again and states that we had a conversation. I said my phone records would prove he was a liar. He said that did not matter. Such integrity. That is correct an officer of the Georgia Army National Guard said that falsifying documents was ok in order to get someone in trouble because that someone reported crime and that the truth did not matter. That is just a little bit of corruption sneaking into the command structure.

What is the point? If it was to prove that people in positions of power can abuse said power to lie about others everyone knows that already. Captain Enriquez tipped his hand by telling me that because i had blown the whistle on crime at the 1-108th I had to be made to look bad. Is that legal? The letter was undated but probably written after June 27th, 2010. At the very end of it carlos enriquez states that i was AWOL and refused to go to "drill". Isn't it strange something like that was written after had blown the whistle on them in Afghanistan and after brian green stated he wanted to punish me?

Nobody had called me on those days, in fact nobody ever called me at all about missing a drill because i never missed one. In fact, there was no attempt by anyone in the 1-108th to contact me at all on those days. No phone calls, no emails, no calls to my next of kin or anyone else on my locater. It is like there was not anywhere I was supposed to be that weekend. The verbiage of the form suggests a drill, so that is what we will go on.

The signature line was the kicker, I lived a half hour from the armory, and nobody ever contacted me at all. If that had been a real drill, they could have driven up to my home. That had been done before other for welfare checks, but it was not because there was no drill. I received the letter in the mail and that pissed me off more than what was written. But once again they want to play, and they were obviously short sighted and did not anticipate an elected official contacting one or two of them years later asking why they lied to a senator and threatened me. At the end of the day there was no drill that weekend. Also, would it not make sense to contact someone that lives a half hour away over that kind of letter being put together?

We spoke face to face at drill. Captain enriquez said that did not matter because i had blown the whistle on the 1-108th so i had to made to look bad "in the long run". That was also when I asked him why randal simmons lied to senator chambliss by telling the senator that I was having an affair with a lady that did not exist. Captain Enriquez refused to answer that question. I do not really blame him; the whole talk was strange anyway because it was obvious, he had a script and was upset that I was not going by it. He even backtracked on the part of the phone conversation about my transfer, now the captain suddenly could not approve it. He wanted me to apologize for reporting crime so he could "help me." That really deserve a fist in the mouth not the honor of wearing captain's bars. But as they say rank has its privileges.

I was supposed to ask forgiveness because I had reported crime after the chain of command refused to do anything! They had no plan beyond the assumption that I would be threatened into submission. He even implied that my complaints going away would make his threat of punishment less likely. When that failed it just became a pattern of lies even to the point of telling lies to multiple elected officials then looping back around to threats. But nothing worked, I would not recant my whistle blower complaint. In their eyes that made me the criminal not the actual criminals who had rank on me. Note that if I had apologized for telling the truth and reporting crime, I was told all the threats would go away. If I did not well...

He also refused to tell me why randall simmons said anyone in the chain of command can forge a signature. That would also be a tough question to answer. Looking back on that and what had transpired I don't look bad in the long run. I could be wrong, but one would think the people that constantly lie about a few issues are really the ones who look bad. He did however show me this little piece of paper.

MAHAN, WESLEY, S. (E5) 256-45-7811 HHT 1-108TH CAV

NAME& GRADE SSN UNIT

Part I. INITIAL ACTION

1. I am considering whether you should receive non-judicial punishment under Section 38-2-360, O.C.G.A. for the following misconduct:

In that you did, on or about 25-27 June, 2010, without authority fail to go, at the time prescribed, to your appointed place of duty to wit: report to drill at Yellow Ribbon Event weekend. This is in violation of § 38-2-519, O.C.G.A.

2. First, I want you to understand that I have not yet made a decision whether or not you will be punished, and I will not impose any punishment unless I am convinced beyond a reasonable doubt that you committed the offense. You may present witnesses and other evidence to show why you shoulde't be punished at all (matters in defense) or why punishment should be very light (matters in extenuation and mitigation). I shall consider everything you present before deciding whether I will impose punishment or the type and amount of punishment I will impose. You are not required to make any statements at all, but if you do, they maybe used against you. ~~If you are a commissioned or warrant officer or were on "active State duty" at the time of the alleged offense(s) you have the right to demand trial by court-martial instead of non-judicial punishment.~~ You do____; do not X ; have such right. I will now give you 24 hours, if you require such time, in which to ~~decide what you want to do if you have the right to demand trial by court-martial and~~ to prepare any matters you wish to present in defense and/or extenuation and mitigation. ~~If you have the right to demand court martial in lieu of non-judicial punishment,~~ I will afford you the opportunity to consult a Judge Advocate before you make your decision. If you desire to consult counsel of your choice and at your expense, I will give you time off from duty to do so; however, you are not entitled by law to representation by counsel in connection with these proceedings.

17 July 2010 CARLOS C. ENRIQUEZ CPT, HHT 1-108th CAV

Date & Time Name, Grade & Organization of Commander Signature of Commander

3. ~~Trial by Court-Martial ____is; ____is not demanded.~~

4. Matters in defense and/or mitigation and extenuation:

____ are not presented; X are attached hereto; ___ will be present in person

17 July 2010 WESLEY S. MAHAN, E5

Date Name & Grade of Service Member Signature

I never found out if he copied this directly or wrote it himself. It was just ironic that it was written to state that enlisted personnel did not have the right to a military lawyer during this process considering I'd already asked for one to be present at that drill.

The 1-108th had this vibe that they could say what did and did not happen even if it was obvious, they were being dishonest. I do not know why this was so, they were used to people just believing what was said or they were used to threatening people into submission. It did not matter to me either way. I only had a few months left barring a transfer approval and the paper trail had already begin to build, we were going to finish this game no matter what and since they had been so kind as to let me know the transfer would not be approved, I could have fun. They did not officially say the transfer would not be approved but when others were while man was sat on it was a big hint.

That is however beside the point. By this time, I had been told that it was not a crime for randall simmons to lie to a senator about me and that forgery and fraud were not crimes. Stack onto that a form about me refusing to go to drill, which the author of the form admits is fake and the "first sgt" threatening me and I am starting to think they do not like people reporting crime. This is all so idiotic what did they think was going to happen? I was not going to be convinced by threats to stay with the 1-108th or recant my complaints. Time was on my side not theirs, sadly officers cannot order time to stop. I am sure the military is working on that. That and the ability to erase time so people do not ask awkward questions. Fort Hood learned.

Kind of like if someone asked why a married male officer and female enlisted member went to a club together in savannah when we were demobing after afghanistan. Why would something like that happen? Let me elaborate on that. When we at Fort Stewart Georgia winding down the deployment four people went to a night club off post. The croup consisted of a male lieutenant colonel, a male Sergeant major, a male sergeant, and a female specialist. The specialist was the designated driver on the way there and the sergeant was the designated driver on the way back. That is a little strange. While we were in afghanistan the specialist came up to the colonel and hugged him in the chow hall. Of course, I could be wrong.

Captain enriquez reiterated his comment that because I had blown the whistle on criminal activity at the 1-108th i had to be made to look bad. This from a man who got his leadership style by copying the captain from band of brothers but ignoring the fact that the captain from that movie treated everyone the same. Captain enriquez had more favorites then a drunk at closing time.

During the durations of our phone call and conversation at drill a pattern emerged. There was the obvious anger at me for reporting crimes but there was a strange undercurrent present. If I would just apologize for reporting crime, they would stop threatening me. Even though there was no internal affairs office in the georgia national guard the 1-108th wanted me to recant. Now what took on place on days? A retreat for married couples. It was not a drill it was two nights of drinking and two days of couples' therapy. It was not mandatory for anyone. There was also this attitude that quoting law at me was supposed to demoralize me. I had reported crime so law would be thrown at me to make me a criminal. So, we have a lack of integrity and no honor.

Chapter 7: Because I am an officer and I said so

While deployed I was dumped, those sorts of things happen. The best part is that I swore that would not happen to me. Life can be like that. However, for some reason that has never been explained to me someone in the of the 1-108th changed my ex-girlfriend's name and held the threat of a court martial over my head. No reason for that was ever given but it was not because I had complained about crimes committed by the click in the 1-108th. Even though there was no other reason unless the 1-108th was just corrupt. Which is it? The threat of criminal charges for having an affair with a woman that did not exist, a woman they made up was presented to me. Obviously, any punishment would be real, but I was told that if I just wrote a statement saying I had an affair in Afghanistan everything would go away. Lie about an affair with the women they made up. They wanted me to lie. I wonder why they wanted me to lie, it was like they did not even have a real charge to throw at me.

I just had to grit my teeth and dig my heels in as we were a few months away from going home. Everything would change when we got back to Georgia. I'd have full access to my bank and insurance statements and information of the legal type. The girl had taken me for some money. But that didn't matter with respect to her it was a paper trail to prove the 1-108th lied. While in afghanistan I was denied access to a printer for non-official business until a friend helped me out. When we got back to the states, I put in some transfer requests which the leadership of the 1-108th sat on. That combined with multiple cases of forgery and them making up a female led to me filing a complaint with the office of Senator Chambliss.

Lt. Col Simmons as you see in these two letters twice told the senator that I had been in a relationship with a married lady that did not exist and in the first letter stated that forgery was not a crime and that anyone could do it! Let us go over the first letter together. But keep in mind the checks i mailed to her from Afghanistan. What checks might those be? She had gotten me to send her some checks because she needed money. The checks were written out to her birth name and signed by her married name. The ones that caused a shift in their story as soon as I was able to print them out. Randall simmons made up his own name for her soon after I was dumped. I never found out why, but I do have an idea. Without further delay, the first letter from randall simmons, whom i understand is now a general and therefore incapable of committing a crime. The first letter around the same time Captain Enriquez wrote the letter in which he accused me of answering the phone and refusing to come in for three straight drill days. I think this was supposed to demoralize me.

DEPARTMENT OF THE ARMY
GEORGIA ARMY NATIONAL GUARD
HEADQUARTERS AND HEADQUARTERS TROOP
1ST SQUADRON 108TH CAVALRY (RSTA)
300 NORTH RIVER STREET, CALHOUN, GA 30701

NGGA-108-CDR 08 July 2010

MEMORANDUM FOR SGT Wesley Scott Mahan, 1020 Peach Road, Cohutta, GA 30710

SUBJECT: Response to Congressional Inquiry

1. This letter discusses two issues relating to your congressional inquiry filed on 24 JUN 2010. First, the issue of your punishment and then, that the wrong NGB Form 4100-1-R-E was filed on your behalf.

2. I have investigated both issues and find the following:

a. Punishment. Mrs. Ivy Ferguson, a married woman, contacted the Squadron's leadership seeking help in stopping unwelcome communication from you. Although maintaining a relationship with a married woman could be considered a violation of several punitive articles of the Uniform Code of Military Justice (UCMJ), no charges for the offense were taken against you. Subsequently, proceedings under Article 15 or Court Martial were not conducted for the offense. Members of your chain of command counseled you about the situation and issued you direct and lawful orders to stop all types of communications with Mrs. Ferguson and documented the counseling on DA Form 4856, (Developmental Counseling Form). This however, does not constitute any type of military punishment. There were two reasons that CPT Enriquez did not want to see any documents pertaining to the situation. First, CPT Enriquez's goal was to stop the communications from you to Mrs. Ferguson and in doing that, the documents that you offered were irrelevant. Second, the situation had the potential of becoming a serious legal matter for you and CPT Enriquez thought that it would not be in your best interest to disclose any evidence that could possibly be used against you at a later date.

b. Wrong NGB Form 4100-1-R-E filed on your behalf. Correspondence from a Soldier to the Brigade Headquarters should first be turned in to the Company/Troop for processing and forwarding to Battalion/Squadron. Battalion or Squadron should then process and forward paperwork to the Brigade level. In all instances mentioned in the inquiry, paperwork was taken straight to Brigade Level for processing without using the proper chain of command or routing procedures. Correspondence handed straight to Brigade cannot be tracked by lower-level units. 1-108th CAV turned in a proper 4100-1-R-E on your behalf. Anyone along the proper channels could have signed a request for the Standby Advisory Board. The most likely scenario is that someone noticed the request for the Standby Advisory Board was not signed so they signed it to make sure you were promoted in a timely manner despite your assignment in Afghanistan. You were promoted based on the 4100-1-R-E and request for Standby Advisory Board and your MOS is not relevant.

NGGA-108-CDR
SUBJECT: Response to Congressional Inquiry

3. For any questions or for more information please contact CPT Geoffrey T. Miller at 706-879-2916 or geoffrey.miller2@gm.ngb.army.mil.

RANDALL V. SIMMONS, JR.
LTC, IN, GAARNG
Commanding

Enclosures
Congressional Inquiry Packet
Email From SFC Teague

“Mrs. Ferguson, a married woman, contacted the squadron’s leadership seeking help in stopping unwanted communication from you.”. Just start off with the lie. He is really sounding certain of that this time. The form he speaks of I never saw or signed, that becomes important later. Now he gets to the good stuff, “the documents that you offered were irrelevant” I must disagree as the 1-108th’s leadership was accusing me of having a relationship with a woman that did not exist. I would say those documents were truly relevant. So right off the bat he lies to the Senator, that does not make anything that could come after that look promising.

Besides the checks there was insurance paperwork which showed her on my auto insurance as her maiden name. The name the 1-108th made up was nowhere to be seen except from the 1-108th. It’s like they had already decided to punish before all the facts were in. Then when all the facts came in the 1-108th decided that telling a lie would be better than admitting they made a mistake. That sounds like an attempt to punish a whistle blower. I am sure they would never do that. He goes on and on about paperwork, but that paperwork was never presented to me. It is like it did not exist.

So, if i am reading his first paragraph correct randall simmons is saying that I was involved with a "mrs ferguson" and that paperwork I had that showed that was all a lie was not important. Now why would he say that paperwork showing the chain of command was willing to lie to punish a service member would be seen as not important? A good commander would have taken the paperwork and stopped the whole thing in its tracks then punished the people that were lying about the service member then repeated the lie to an elected official. But what do I know i was just an enlisted member. Randall Simmons mentions carlos enriquez wanting to avoid legal issues and that was why he refused to allow me to present the paperwork. In a way he was telling the truth then.

Whose legal issues? Why would carlos enriquez refuse to let me send paperwork to the 48th brigade command that shows the 1-108th command was accusing me of having an affair with a female that the 1-108th command made up? What would possess him to spike that information? Why would he suppress information that showed that someone in the chain of command had written a false statement about me and that charges based on those false statements had been held over my head? How is that helping me at all? Who is that helping?

There is only one possible or logical choice. The chain of command was not worried about legal action against me, they were worried about legal action against them! That is why I was never shown the counseling statement or any of the other "papers" and probably also why they were putting pressure on me to write my own statement about "mrs ferguson". That is so devious. It is also stupid. I'm sure he was very worried about legal issues hence the pressure to write a statement before we rotated back to the states. But I did not and now they had a dilemma. Off active duty, on drill status, now really pissed off, this was not the situation they thought we would all be in at this point. I was supposed to write a false statement for them while we were in Afghanistan. When I did not, they lost their plot.

This was also very disturbing because randall simmons rationale is that evidence I wanted to show carlos enriquez could have been used against me. So, he is saying that bank and insurance paperwork that showed that the women had taken me for money then dumped me could be held against me? How? It also showed that the 1-108th was pushing me to lie. In addition to that the paperwork showed that I had been dumped by someone in the United States while I was in Afghanistan. Carlos Enriquez was attempting to get me to write a statement saying that I had been having an affair with a woman in afghanistan that the 1-108th had made up. So, if he had taken those papers the lie the 1-108th was pushing would have been exposed. Legal issues indeed.

So, the legal issues they were worried about were not mine, the issues were the leadership's issues. Now does that not make it all better or does that make it worse? They were refusing to let me provide evidence that showed i was innocent of what they were accusing me of but why? Why would a Captain in the Georgia Army National Guard refuse to let a service member under his command present evidence that showed the service member was innocent of the crimes the captain was accusing the service member of committing? What would the captain gain by suppressing evidence that showed the service member was innocent of the crimes the captain was accusing the member of? Why would the captain do that? That was yet another thing they did not want to talk about when we got back to Georgia. When i was at a small outpost on the Afghanistan/Pakistan border, they could not stop talking about these things but not when we got back to Georgia. Then it would become their goal to talk about anything but that. Even if that meant telling as many lies as possible.

I was on Torkham Firebase on the Afghanistan/Pakistan border when all that happened. The base was very small, there were no support offices of any kind. In this case that included no jag office. So, when carlos enriquez, brian green or anyone else that had talked to me would threaten me with charges over a made-up lady I had nobody to talk to in a legal way. Now take that situation and add carlos enriquez suppressing evidence. I had no way to get to a jag lawyer without going to another fob which I could not do without permission from the people suppressing the evidence! That meant it was gut check time. For a bunch of Georgia Army National Guard leaders, they sure did not have a grasp of tactics as all I had to do is dig in my heels and grind out a few months and we were back in the United States. Patience is a virtue.

Also, this letter had a sinister element to it. Prior to this inquiry I had complained to the chain of command about crime taking place at the 1-108th and received threats as a result. That was the main reason for the congressional inquiry. Their response was to say that forgery was not a crime and to tell the senator that I was a criminal not a whistle blower. If only there was some sort of law that made it a crime to lie about and issue threats to a whistle blower.

The second paragraph. Randall Simmons is correct in stating that paperwork going to brigade level should be turned in at the unit first. However, in this instance I had in my hands physical proof in the form of paperwork from Atlanta that showed at least two times in which paperwork that i turned into the 1-108th was thrown away and forged paperwork was sent down instead. So, after seeing them toss my documents and forge replacements they lost the right of any more chances. I said that before I sent in the correct documents. I taunted them, perhaps that had something to do with what happened later.

The administrative office was in the building next to the communications office when we were in Afghanistan. Maybe they could have told me? And in that particular instant it was known that the promotion paperwork was being sent in outside of the administration section. They had messed up too many times. I told them that they had failed to many times, and I would take the paperwork in myself. There was backlash and anger from that, but I did not care. Since administration could not do their job, I would.

The results of those two (at least) times impacted negatively on me. "1-108th CAV turned in a proper 4100-1-R-E on your behalf.", no they did not. Not only does he tell the senator's office that i was involved with a female that did not exist, but he also lies about the paperwork. That is sickening and is not the behavior one would expect of an officer of the Georgia Army National Guard. No integrity or shame is present in randall simmons. He also fails to mention that the paperwork being thrown away and replaced with forgeries. He also lies about the process. The standby form was needed because I was listed under the wrong MOS, no form, no promotion. I was listed under the wrong MOS because the admin section of the 1-108th threw away the form I turned in and forged one on their own.

Sadly, his shame would not kick in until much later and even then, only for one thing, in the meantime he rolls on. The standby advisory board paperwork is to be signed by the person that was put into the wrong MOS slot. That person must sign it, not a Sergeant, officer, or anyone else. That person. I filled one out, signed it and turned it in myself. As the 1-108th had already thrown at least two documents id turned in at that point i wanted to make sure it got to where it needed to be, no more magic tricks. So, if his last sentence is read right a from filled out incorrectly and not signed and another from filled out correctly with a forged signature on it were responsible for me getting promoted. Not the form I filled out correctly, signed and then took to the brigade. Their dishonesty did get a little trying at times. Of course, he also endorsed forgery as randall simmons stated anyone in the chain of command can sign a document for me.

Also, it would be remiss of me to not mention that did punish me for missing a yellow ribbon event that I was not supposed to be at. It was a couple's retreat. It was a choice they said we could go for an extra paycheck since we did not have to be a drill yet, lots of people did not go. Aside from me nobody that did not go was punished. But those people did not blow the whistle on crime and stand up to threats from the chain of command. During the conversation in which carlos enriquez defended the letter and the lies he also indicated he was going to file paperwork to punish me for missing a drill. He cited O. C. G. A. 38-2-360. Here it is.

O.C.G.A. 38-2-360 (2010)

38-2-360. Commanding officer's nonjudicial punishment; regulations limiting; officers in charge; appeal; effect on more serious offenses

(a) Under such regulations as may be issued pursuant to this chapter, any commanding officer, in addition to, or in lieu of, admonition or reprimand, may impose one of the following disciplinary punishments for minor offenses without the intervention of a court-martial:

(1) Upon officers and warrant officers of his or her command:

(A) Withholding of privileges for a period not to exceed two consecutive weeks;

(B) Restriction to certain specified limits, with or without suspension from duty, for a period not to exceed two consecutive weeks; or

(C) If imposed by the Governor or by a general officer, fine or forfeiture of not more than one-half of one month's pay per month for two months; and

(2) Upon other military personnel of his or her command:

(A) Withholding of privileges for a period not to exceed two consecutive weeks;

(B) Restriction to certain specified limits, with or without suspension from duty, for a period not to exceed two consecutive weeks;

(C) Extra duties for a period not to exceed two consecutive weeks and not to exceed two hours per day, holidays included;

(D) Reduction to the next inferior grade if the grade from which demoted was established by the command or an equivalent or lower command;

(E) If imposed upon a person attached to or embarked on a vessel, confinement for a period not to exceed seven consecutive days; or

(F) If imposed by an officer exercising special court-martial jurisdiction, fine or forfeiture of not more than one-half of one month's pay per month for two months.

(b) Under such regulations as may be issued pursuant to this chapter, limitations may be placed on the powers granted by this Code section with respect to the kind and amount of punishment authorized, the categories of commanding officers authorized to exercise the powers, and the applicability of this Code section to an accused on active state duty who demands trial by a court-martial.

(c) An officer in charge, for minor offenses, may impose on enlisted persons assigned to the unit of which he is in charge, such of the punishments authorized to be imposed by commanding officers as may be specifically prescribed by regulations issued pursuant to this chapter.

(d) A person punished under authority of this Code section who deems his punishment unjust or disproportionate to the offense may appeal, through the proper channel, to the next superior authority. The appeal shall be promptly forwarded and decided, but the person punished may be required in the meantime to undergo the punishment adjudged. The officer who imposes the punishment, his successor in command, and superior authority shall have power to suspend, set aside, or remit any part or amount of the punishment and to restore all rights, privileges, and property affected.

(e) The imposition and enforcement of disciplinary punishment under authority of this Code section for any act or omission shall not be a bar to trial by court-martial for a serious crime or offense growing out of the same act or omission and which is not properly punishable under this Code section. However, the fact that a disciplinary punishment has been enforced may be shown by the accused upon the trial by court-martial and, when so shown, shall be considered in determining the measure of punishment to be adjudged in the event of a finding of guilty.

(f) Whenever a punishment of forfeiture of pay and allowances is imposed as provided in this Code section, the forfeiture may apply to pay or allowances becoming due on or after the date the punishment is imposed and to any pay and allowances accrued before such date.

What do we not see in that statute? Forced mental health evaluation is not there in fact there are steps that are supposed to be taken if one of those is to happen. We will get to that later. Being mad at someone under your command because that person reported crime is not a valid reason. That did not matter the only thing that mattered was that someway anyway they could the 1-108th had to make me look bad. But on my last drill? When they had no intention of holding a real hearing or giving me the hearing, I demanded? The statute was not a bar to a court martial. Prior to this I had requested a lawyer to be present at drill. That request was never honored, it could not be because the jag lawyer would have been put in the situation of having to investigate high ranking individuals or just to ignore the crimes. Surely a military unit would never suppress information to protect people of high rank would they? Fort Hood.

So, during the July 2010 drill weekend carlos enriquez starts the ball rolling on my punishment. Despite my request for a jag lawyer to be present at drill carlos decreed that i would have no access to a jag lawyer. So the august drill rolls around and there is nothing done on that, nobody spoke to me or gave me any paperwork regarding the hearing. So, heading into the September dill, which was my last drill, there was going to be the hearing. To the shock of nobody my request for a jag lawyer was denied again and I was told I would lose a stripe which I appealed. But the real gem was the bit tacked onto it in addition to the punishment. Also, the hearing was me representing myself in a room with the entire chain of command, minus brian green, and a few other high rankers that wanted to watch apparently. It was a show after all. It was never about a yellow ribbon event it was about punishing a whistle blower. Here is randall simmons choice:

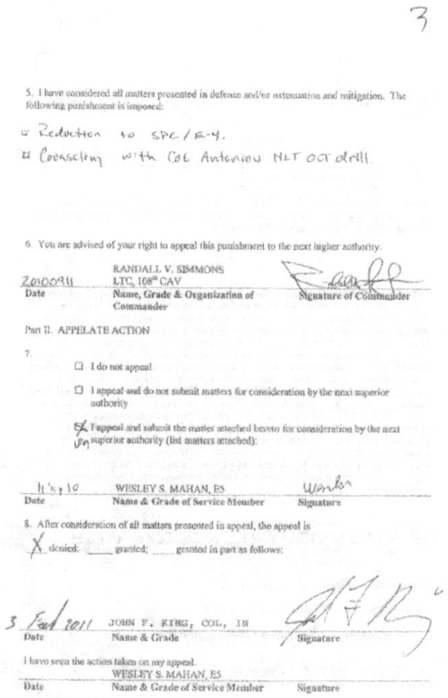
3

5. I have considered all matters presented in defense and/or extenuation and mitigation. The following punishment is imposed:

- Reduction to SPC / E-4.
- Counseling with Col Anteniou NLT oct drill

6. You are advised of your right to appeal this punishment to the next higher authority.

20100911	RANDALL V. SIMMONS LTC, 108th CAV	
Date	Name, Grade & Organization of Commander	Signature of Commander

Part II. APPELATE ACTION

7.

☐ I do not appeal

☐ I appeal and do not submit matters for consideration by the next superior authority

☒ I appeal and submit the matter attached hereto for consideration by the next superior authority (list matters attached):

11 Sep 10	WESLEY S. MAHAN, E5	
Date	Name & Grade of Service Member	Signature

8. After consideration of all matters presented in appeal, the appeal is

X denied; ____ granted; ____ granted in part as follows:

3 Feb 2011	JOHN F. KING, COL, IN	
Date	Name & Grade	Signature

I have seen the action taken on my appeal.

	WESLEY S. MAHAN, E5	
Date	Name & Grade of Service Member	Signature

"Counseling with col anteniou nlt oct drill", Randall Simmons knew perfectly well there was going to be no October drill for me unless of course one of my multiple transfer requests were approved. In that case I would not be there, I might not even be in the state. While he was presenting that he told me he would not sign any transfer requests. But counseling? Well, this was sold to me as "legal advice" and it was already scheduled for the next day that is the Sunday of my last drill.

Whatever right? I drive down to Atlanta to the appointed place in the Georgia National Guard command facility and end up in the parking lot of the mental health building. This was not a jag lawyer's office. But I went with it, I was there after all, and this could be fun. I had brought all the paperwork about the crimes with me including randall simmons's letter to the senator and the check stubs and insurance paperwork that showed what really happened. I presented my case. When I was done the officer was almost speechless. What did randall simmons think I was going to do? This was a huge tactical blunder on his part unless he was sure the officer, I spoke to would say and do absolutely nothing, maybe he was right.

Randall Simmons just broke the law; he broke a few laws. He used non judicial punishment to force a service member into a mental health evaluation. Just to make me look bad, they knew I was fine, but they had to "get me" for blowing the whistle on crime. He knew there was no justification for that. He just showed his desire to punish a whistle blower. What a wonderful leader, I am sure fort hood would love him. For some reason, the Georgia National Guard would do anything up to telling more lies and threatening me when I asked about that punishment later.

Now why would randall simmons put illegal punishment onto a form he pushes onto me on my last drill? That doesn't exactly scream integrity. I did not miss a drill and the entire thing was an attempt to make me look bad because I blew the whistle on criminal activity at the 1-108th. At this meeting I mentioned the forgery and threats but was immediately interrupted by joey recker and told to let randall simmons speak. Does it even matter? This was my last drill. It turned them on, all that power, all that ability to lie about anyone.

This also exposed a huge flaw in randall simmons' critical thinking. Yes, they "got me" in their mind but what took place? I was out and the crimes were still there but their ability to threaten me was not. I did have other things that were of more importance, but they just gave me a project. Of all the service members to mess with why would they choose the one studying criminal justice and psychology in college? I wonder what john king was told when he signed off on the hearing months after I was out.

Realistically I did not expect the Georgia National Guard to do the legal thing, but I could have fun with it and who knows what could happen? It is not like I would catalog everything and author a book about it. Also given the Georgia National Guard's penchant for not being totally accurate with paperwork this would only work for them if I just rolled over and died. He also exposed himself. That punishment was illegal, sure it made people feel better, but it was illegal.

Chapter 8: What he meant to say was

In 2011 I asked Senator Chambliss' office to contact randall simmons again and they did. The extra time to think about it did not do randall simmons any good. He still held onto the lie, but he did amend this version of what took place a little bit regarding forgery and fraud. He amended it quite a bit. At this point I imagine randall was moving up the chain of command and getting his picture in all the nice little public affairs write ups. He did not want to be bothered about crimes he allowed and committed.

6

DEPARTMENT OF THE ARMY
GEORGIA ARMY NATIONAL GUARD
HEADQUARTERS, 48TH INFANTRY BRIGADE COMBAT TEAM
1030 SHURLING DRIVE
MACON, GEORGIA 31211

NGGA-BCT-DCO 15 May 2011

MEMORANDUM FOR Wesley Mahan, 1020 Peach Road, Cohutta, GA 30710

SUBJECT: Response to Second Congressional Inquiry

1. This correspondence is in response to your inquiry filed with the Honorable Saxby Chambliss' office on 15 March 2011. Specifically, this memorandum will address claims of forgery and punishment. In addition, I will attach a copy of the congressional response from 08 July 2010 where some of your concerns have already been addressed.

2. I have investigated these allegations (again) and render the following as my final response.

a. Forgery. The 1-108th Cavalry Regiment and 48th Infantry Brigade Combat Team does not condone forgery for any reason. You allege that your EPS Worksheet and Standby Advisory Board packet was forged. As our unit moved from Ft. McLellan, AL back to Georgia and then to Ft. Polk, LA, during the January-February 2009 timeframe, it is apparent that some of your paperwork was misplaced or lost. As a result, the unit recreated the paperwork from electronic records and submitted it on your behalf. You claim the signatures on these documents are forgeries. I cannot confirm or deny that at this point, but I can certainly confirm that, if done, it was done with only your best interest in mind. In fact, the submission of this paperwork led to your promotion to Sergeant while deployed later in 2009; this would not have happened without the documents you mention. If you wish to challenge these documents, our only recourse would be to revoke your promotion order to Sergeant. However, considering that you are no longer a member of the Georgia Army National Guard that would not be prudent at this point. Your claims of forgery have been investigated and ruled to be unfounded as there is no evidence that these records were actually forged or that you suffered because of them being submitted.

b. Punishment. Please see the first rebuttal dated 08 July 2010 where I addressed your concerns from the counseling directing you to stop contacting Mrs. Ferguson. As a reminder, counseling statements are administrative documents and, in this case, simply designed to formally record what was said. There was no punishment associated with that event.
Your recent inquiry claims that you were wrongfully punished for missing drill in June 2010. You also claim, via e-mail, that since it was a Yellow Ribbon event that it was not a "real" drill. That is not the case. The drill period for the 1-108th Cavalry for June 2010 was a Yellow Ribbon Reintegration event conducted in Atlanta, GA 25-27 June 2010. All Soldiers received travel orders to attend. Based on numerous emails you sent prior to the event you acknowledged the event and expressed concern for your safety. You were never able to articulate why you were afraid to attend and as such, you were not granted leave. There was certainly nothing to jeopardize your safety at the event. Your commander requested that you receive non-judicial punishment for being AWOL. I heard your case and decided to reduce you one grade to

7

NGGA-BCT-DCO
SUBJECT: Response to Second Congressional Inquiry

Specialist/E-4. You appealed that decision and COL John King, 48th IBCT commander, denied the appeal after reading the file. That action is closed.

3. I am the point of contact for this matter. Please consider this my final response to your inquiries.

RANDALL V. SIMMONS, JR.
LTC, IN, GAARNG
Deputy Commander

2

It starts with, "...does not condone forgery for any reason.". It takes a special kind of immorality to go from defending forgery to saying you do not defend it just because you might get in trouble. Look at the documents, read his first letter to the senator, what the hell randall? "...it is apparent that some of your paperwork was misplaced or lost.". It is entertaining to see someone trying to cover their ass that obviously has never had to, one of the documents was forged in March but he had to say something, and the truth was out of the question. The truth would have gotten him in trouble or hurt his image. His image was important.

He goes on, "...claim the signatures on these documents are forgeries. I cannot confirm or deny that at this point.", I can, the documents were forgeries. But randall is an officer and the leadership culture requires him to lie to protect himself. He then goes on to say that if any forgery occurred it was, "... your best interest in mind.". Me getting massively underpaid on travel vouchers was not in my best interests. My pay being short for an entire year with no back pay ever was not in my best interests. Randall simmons an officer in the Georgia Army National Guard just protected criminals and promoted crime. Then he lies again as he had previously implied that the forgery was my fault because I took the paperwork straight to brigade after the latest round of forgery. Now he says that the forgery was the reason I got promoted! Obviously, a scare tactic but a shallow one at least. They would have to rescind my promotion! I was already out! How does a person go from saying a crime cannot be confirmed or denied to saying the crime is responsible for an action having taken place? This man is an officer? A leader? He confirms what he cannot confirm? This kind of dishonesty is how the situation at Fort Hood was created.

A moment on randall simmons claim that forgery got me promoted. He chides me for going straight to brigade level with the correct paperwork because the forgery got me promoted. How does that make sense? He is saying that someone in Calhoun that forged a document and signature then sent it to Atlanta, while I was in Afghanistan, has more control over me then I do. Does that sound legal? That someone in Calhoun can forge the signature of someone in Afghanistan and therefore control what happens to the person in Afghanistan?

The paper I filled out in Afghanistan counted, the one they forged did not. That was touched on during his first letter. Feeling a lack of respect from randall simmons here it's almost like he did not actually write the first letter. To the best of my knowledge when the brigade rotated back to the states the files from Afghanistan were combined with the files in Atlanta. I wonder what they thought when they saw mine if they cared at all?

The arrogance here is out of control. I am one person but according to the letter randall simmons sent to the senator anyone in the Georgia National Guard is at the mercy of the administrative section. How so you might ask? Well multiple my case by whatever number you think of those people would be at the mercy of the administration section in Georgia even if they were serving overseas. According to the example laid out by randall simmons in his letter to the senator the administration sections have complete control over the military members! They can forge documents and signatures and a flag officer will let it happen! That is something no recruiter ever told me.

Then, "your claims of forgery have been investigated and ruled to be unfounded". Earlier in the paragraph he said he could not comment on forgery now he is saying that it never happened. It is the logical next step for him, first he denied that it happened, then threaten, then lie then back to deny. What is the logic here? I sent copies of the forged documents in many times it was not a secret. This is what happens when a military unit has no internal affairs section. This is also an insult to the senator; how can something be unfounded if you just said that it happened? He still will not speak of that travel voucher. That would have made his pattern of lies a little more difficult to pull off.

But wait! I'm getting multiple vibes from randall about forgery. One is that there was none which makes sense to say if one is defending a reputation or organization. There was no forgery but there was, but it was not real, but it was. There might have been some, but it was only with good intention? Is that what they tell people who have pay errors and end up on the wrong list for promotions? Lastly or firstly in the letter there is of course the promise that he would never condone forgery even though in the first letter he did. It must be nice to be an officer.

Of course, he implies that I was violating the UCMJ, "...directing you to stop contacting Mrs. Ferguson..." Remember that name because he is going to chuck it out window soon. I was told on paperwork and in person, to break contact with Mrs. Ferguson. That is what randall simmons is saying. He mentions the counseling statement again that I had never seen but starts to dial back the rhetoric. Perhaps he realized that at this point I was out, and he was talking to private citizen not a sergeant that he had rank on. Or perhaps he had some shame in him after all this is the second time that he has willingly lied to senator Chambliss. I wonder if he was nervous or if he was worried about the lie being exposed? Obviously not because he would have people that would lie for him then threaten me when the lies went nowhere. True leadership.

He speaks of the yellow ribbon event and states that he saw no reason for me to fear for my safety from anyone in the 1-108th. So, the first sergeant telling most of the chain of command that he wanted to punish me for reporting crime is not something to be concerned about? Them sitting on a transfer request while at the same time holding a lie over me is, ok? There was this push by the 1-108th to make physical harm the issue. That was never the issue, the issue is the 1-108th is ok with forgery, fraud, and threatening whistle blowers. Not my kind of place.

A few simple facts. Since I discovered my signature had been forged a few times by the 1-108th things had gotten interesting. I'd been threatened a few times, with one of them being an email by the top enlisted member of the unit stating he wanted to punish me. I've had two transfer requests sat on for unknown reasons and had been told that forgery and fraud were just the way the 1-108th operated. Now obviously randall simmons cannot mention that because he would incriminate himself due to his actions and inactions, so the "physical harm" lie was born.

That was pushed by a few people in order to attempt to get me to stop blowing the whistle. It did not work. The ease at which randal simmons creates the main lie to the senator is very troubling. This man, this officer in the Georgia National Guard thinks nothing of repeating a lie designed to make me look like a criminal. As he protects real criminals, he lies to an elected official saying I am a criminal. What leadership. Is this what the Georgia National Guard sees as command material? Sadly, it would appear so.

The physical harm issue was an attempt to hide what was being said and done. When I said that I did not want to be there because the leadership was fine with document and signature forgery there was a reaction. The reaction was not to stop breaking the law it was to threaten the person reporting the crime. That is illegal in so many words. That is also not a place i want to be, the idea that randal simmons is comfortable in that kind of environment says many things about his ideas for leadership.

If a unit can forge documents and signatures without fear of the law, then why would anyone feel safe in that environment? So instead of punishing illegal behavior the leadership of the 1-

108th choose to call the whistle blower a coward that thought people wanted to fight him. This was done even though the proof had already been sent to the chain of command.

He claims that forgery cannot be confirmed or denied then says it was investigated, in the same letter he says this. That is a weak investigation. Did he even read what he wrote before he sent it in? He wrote that knowing that I had copies of documents where my signature had been forged or misrepresented, lies to the senator without even missing a beat. That is not a good thing. Is this leadership? is this integrity? What is going on in the Georgia National Guard?

Also as mentioned before the punishment was not legal. Part of this might have been that they just expected me to go away, I mean that is probably what all the other people they fucked over did so why wouldn't one more? Why else would all that have taken place with such ease unless they were used to doing it? What are the odds of someone digging their heels in and keeping this alive for years? Maybe if the service member was studying criminal justice and psychology in college? Maybe that service member would see all this as a project? Just a thought. But let us dig into that punishment.

Now I was not told when or where this hearing would be, nor was I given a chance to defend myself or bring evidence to the table. Randall Simmons claims that John King reviewed a file and decided. So instead of letting me go to the hearing to face my accusers John King read a file. A file from a unit that forged documents and my signature multiple times. I wonder what that file said. Is that constitutional? Instead of giving the accused a chance to rebut the charges a file is written and given to the individual acting as judge? A file written by criminals is considered evidence. What integrity. John King had no problem signing that form knowing I had not and would not be given the chance to defend myself.

Also read those two letters again, do they read like they were written by two different people? Any differences in structure or content stand out? The second letter shows a lack of knowledge about the promotion process while the first letter reads like someone describing what they did. The 4100-1-R-E and the standby document are totally different forms. The standby document is only sent in if there is a problem with the 4100-1-R-E. They are not sent in together. Again, they are separate documents, and the standby is used if the 4100 gets lost or has a problem. If the 4100 is correct the standby form is never sent in. How did someone seeing something unsigned become all my paperwork being lost then recreated electronically? And why call a senator "honorable" when you are lying to the senator? It is like the first letter and the second letter were written by different people. Why else would the second letter blatantly contradict the first one? In fact, I will just throw it out there that it is possible that randall simmons did not write the first letter. They do read differently after all.

He speaks of emails I sent, and it just so happens i have some of those that I managed to save, and we will see them later. But first the email exchange between myself and randall simmons. This was after he wrote the second letter. I honestly did not think he would reply. But it was worth a shot and maybe, just maybe he might slip up and be honest. The fact that it would be a mistake on his part to be honest shows how sad this situation is. Sadly, due to a failure to print an extra page of emails before my Dalton State College account got deleted post-graduation, I only have a sliver of my first email to randall simmons.

I do not have much left of the email but i have his reply, my reply to his reply and his reply to my reply of his reply. Basically, asking him why he lied to the senator about me. Now keep in mind this was after I was out and sent from a Dalton State College email account. There is no way he would reply or if he did reply admit to the lie, right? I mean that would just be stupid.

Printable Page 2 of 3

8 of 8

Date: Saturday, July 9, 2011 10:32 pm
Subject: RE: who is ivy ferguson? (UNCLASSIFIED)
To: wmahan01@daltonstate.edu

> Classification: UNCLASSIFIED
> Caveats: FOUO
>
> Mr. Mahan,
>
> You are correct. The Mrs. Fersuson mentioned in my responses to
> you is in
> fact [redacted] That was an oversight, but I think you
> figured out
> who we were talking about. Please see the attached email
> traffic between
> [redacted] Your
> relationship was deemed to be inappropriate and we were asked to
> have you
> stop communicating with her. We did that and documented the
> conversationwith an administrative counseling statement. The
> record on this is clear.
> You were in no way punished for communicating with [redacted]
> After you
> were instructed not to communicate with her we received no more
> complaints.Case closed.
>
> As for your claims of forgery, please attached inquiry response
> para 2.a. I
> consider this case closed as well.
>
> The only punishment you received was for being AWOL to the
> Yellow Ribbon
> event 26-27 June 2010. I have all of the email traffic between
> yourself and
> SFC McStotts showing that you knew about drill and failed to
> attend. Your
> punishment was administered and you appealed my decision.
> However, the BDE
> CDR denied your appeal. Case closed.
>
> I see no further reason to continue these discussions. Best of
> luck in
> school and with your future endeavors.
>
> r/
>
> LTC Simmons
>
> LTC Randall Simmons
> 48th IBCT
> Deputy Commander
>
>
>
> -----Original Message-----
> From: wmahan01@daltonstate.edu [mailto:wmahan01@daltonstate.edu]
> Sent: Saturday, July 09, 2011 1:44 PM
> To: Simmons, Randall V Jr. LTC USA NGGA 48th IBCT
> Subject: who is ivy ferguson?
> Importance: High
>
> Mr. Simmons;
>
> As the subject line attests this is a direct question that you
> have so far

they changed her last name so they could still threaten me with an adultery charge

http://dscmail.daltonstate.edu:6777/...

I think the hidden words would be obvious, but I will explain them anyway. That girl had nothing to do with any of the threats, acts of forgery, fraud, or otherwise criminal acts the 1-108th committed. She had nothing to do with the 1-108th leadership telling me to write a false statement saying I had an affair while in Afghanistan. So why put her name out there? Especially when writing about criminals. The willingness of randall simmons to lie about her as well as lie about me was very troubling. Why give amyone another chance by putting her name out?

An oversight? They wanted me to write a false statement so they could punish me based on a lie and it was an oversight? What leadership. So since i was no longer in did he have no fear of admitting he lied to a senator? I mentioned the hearing to misdirect him in the hope that he would give up the name of whoever else made that name up. I do not think randall simmons made the name up personally, but he did "own" it now. While not telling me who did it admitting that it was made up was even better. To this credit he showed zero shame, it is not ferguson anymore (after I remind him) and this time all the email traffic and documents show that except that he did not attach them, yes, he said he did but he did not. Probably because they did not exist. His mention of the counseling statement,which I was never shown, is even more damning. Now it does not say ferguson or at least that is his claim. It was easy for him to change the name twice because the 1-108th consistently refused to put the paperwork or emails in front of me, ever. That does not exactly scream integrity.

He also ducks the draw when it comes to the forgery. Just refers me to his previous letter(s)? If it was supposed to be attached to that email it was not. But given randall's lack of respect for the law it did not matter. I was just supposed to agree with whatever they said. I replied. He replied to me by saying he would never reply to me again. How brave of him. He honestly thought that was it and he would never again hear about or from an enlisted member that he had broken the law to punish.

What does one say to that? What a coward. Yet his cowardice and criminal behavior is protected by the military. All the bluster, all the threats, and he just hides. This gave me the opening I needed to contact the Georgia National Guard IG, now that I had a confession from randall simmons they would act, right? To break that down, randall simmons told a United States senator that a whistle blower was a criminal. Then when that did not shut the whistle blower up the next step is an illegal punishment designed to destroy the character of the whistle blower. Randall simmons also uses his position to destroy the career of the whistle blower. What a man of integrity randall simmons is. Think about it. He had no problem lying to a senator to protect criminal sand punish a whistle blower. It was like second nature he did not even have to think about it. When he gets caught, he just switches the names and events around and why not? He knows that the military does not like to punish officers.

Chapter 9: Noise just noise

Since the IG office is the closest thing the Georgia National Guard has to an internal affairs office that was who I had to contact. I already knew that the IG office had zero interest in punishing an officer. I got a taste of that when I was still with the 1-108th. The IG office would listen to me and say that crimes were going on in the 1-108th but as soon as they heard the names and ranks of the criminals, they refused to do anything. But I thought they needed to be reminded of what their job really was. If there was any hope of my going back in after graduating college, I had to deal with this.

That is a shame because randall simmons changed his story with absolutely no shame even though that meant admitting that he lied to a senator. His rationale about the hearing is worse, he had no intention of me going to that hearing which was why the process was started on the Saturday of my last drill. A file? A file written by people who have forged signatures and documents. Does that not sound a little strange or wrong?

How can he even write that? What is the legality and morality of a national guard unit writing a file and sending it to a hearing instead of telling the (now former) service member when the hearing is? I had requested a lawyer and a hearing, both those requests were ignored, and my signature was probably forged again since I was already out when they had the hearing. Or they could just leave it blank.

At this point I had been out for a bit and was studying criminal justice and psychology at Dalton State College. This had become a project for me, was there a point given the evidence that the Georgia National Guard would admit the 1-108th broke the law and deal with it or would they regress to threats? They could not keep lying in letters because I kept every letter and even those started to contradict each other. Eventually they would have to either tell the truth or resort to threats. Spoiler alert, they resort to threats.

Specifically, will randall simmons use his authority as on officer to order people to lie for him or would he admit to his crimes? Is there a tipping point? At what point will someone in the Georgia National Guard realize that honesty is the best policy, and will anyone be punished? Obviously, I underestimated the lack of integrity in the Georgia National Guard.

This is not an incident isolated to one branch or one command, before we continue let us look at Thomas Fiscus from the United States Air Force. Thomas Fiscus was a two-star general who at the height of his career was the general over the Air Force's JAG Corps. However, in 2004 it was discovered that he had been having affairs with at least 13 women over the previous ten years. (he was married at the time) His punishment was that he was allowed to retire as a colonel.

Contrast that to Jennifer Jones an instructor at Sheppard Air Force Base who was fined and reduced in rank for having a relationship with a former but possibly current student at Sheppard Air Force Base. Her defense attorney mentioned the Thomas Fiscus case, but that line of reason was ignored. The apparent logic being that one "relationship" (relationship defined by the ability to punish in this case) is worse than at least 13 affairs.

Now contrast these two cases, one with multiple cases of adultery the other a case of two adults boning or not boning I do not know. Which incident deserved more punishment? Is it worse to have a relationship with a possibly current student with both adults being single or is it worse for a married military lawyer to have at least 13 affairs over a ten-year period? Which of those cases had the greatest negative impact from the crimes committed? How many lives were impacted or ruined by Thomas Fiscus's criminal actions? At least 26 and for that he was allowed to retire as a colonel. One would think that multiple cases of a crime would result in punishment. I wonder if any female that reported Thomas Fiscus was ever punished like the writer of this book's foreword was for reporting sexual harassment? Oh well back to reality.

Anyway, I contacted the Georgia National Guard IG in Atlanta and the National Guard IG in Virginia. The IG from Virginia responded first but they just let me know they would send everything down to the state IG in Atlanta. Kicking the can down the road is always a good thing. It was just a coincidence that the complaint regards an officer intentionally telling lies to a senator in order to destroy the character of a whistle blower. Read on.

NATIONAL GUARD BUREAU
111 SOUTH GEORGE MASON DRIVE, AH2
ARLINGTON, VA 22204-1382

October 3, 2011

Office of the Inspector General

Mr. Wesley Mahan
P.O. Box 125
Cohutta, GA 30710

Dear Mr. Mahan:

The NGB IG office has done a preliminary analysis of your request for assistance concerning forgery, removal of false statement and 1SG threat to end your career. You have already addressed most of these issues with the GA IG office. Also you have received two Congressional responses from your GA chain of command addressing the same issues. We have forwarded your complaint to GA IG office to ensure all of your issues were addressed. Be advised that the GA IG office will not address any issue which were addressed through the previously mentioned channels unless the GA IG feels further clarification is needed. The final response will come from the GA IG office.

The point of contact for your case in the GA IG office is MAJ Richard Song. He can be reached at 678-569-6027. This office will take no further action since your issue is being addressed by the GA IG office.

Sincerely,

Michael D. Green
Major, U.S. Army
Chief, Assistance Division

The letter from major michael D. green does little to shed any light on why randal simmons lied to a senator about me and covered up forgery and fraud. The major does an incredible job of addressing the issue without addressing the issue! A true officer! He states that the NGB IG has reviewed the issues but then says they will not address any of the issues! What was the point of even writing a letter if you are not going to do anything? A more troubling aspect i that they reviewed multiple cases of forgery, lying to a senator and a due process violation, and decided there was nothing to see there. The military does get like that sadly. Officers do not like to punish officers. It is a class system. It was also too easy to kick the whole problem back to the state level.

Major michael D. green points out correctly that I received two responses to congressional inquiries, but he does not mention that neither inquiry had its questions answered. At least answered in a way that was not illegal. As has been covered in the first inquiry's response randall simmons said that forgery is not a crime, and anyone can do it. Why would a major in the United States Army do nothing when shown that an officer of the Georgia Army National Guard lied to a senator and covered up crimes? I mean one would think an inspector general officer would jump on a case like that, especially since the crime and confession were given to him in randall Simmons' own letters and emails. But he did not. Why would that be since I sent them all the paperwork I had, and it showed the crimes and the cover up.

So, "removal of a false statement" that part of the letter immediately lets me know that major michael d green of the United States Army had no intention of investigating anything. I didn't complain about false documents being taken out of my record but being put in! To make it appear that he's doing something major michael d green states that he transferred the issues to the Georgia National Guard IG office, the same office that has already refused to do anything. Now that takes willful ignorance, how many other crimes did he transfer away? Does it even matter? Fort Hood.

Major michael d green then states that the Georgia National Guard IG office will not address any issues that had been previously addressed through any channel the major had mentioned. So, following that logic nothing will be done. Contact information is given which made no sense because major michael d green knew that I had already contacted them. Unless the two offices had already spoken and decided to not do anything. Surely that would never happen. He continues the pattern of ignoring forgery, fraud and lies to elected officials. Major michael d green also did not want to talk about the email from brian green for some reason. No reason was ever given to me, but that email was treated like poison more so than randall's letters and emails.

This letter continued the unfortunate standard that the Georgia National Guard gladly upheld. I would contact a representative from the Georgia National Guard, show them the letters, emails and documents then listen as the representative would say it was horrible and they would investigate it. Then if a reply was sent it would be a letter that would either lie or just state they were not doing anything. But they had to know I was keeping copies of everything right? Why would they intentionally do nothing when it is reported to them that an officer of the Georgia National Guard lied to a senator to punish a whistle blower?

So how is kicking it back down to the very level that refused to anything solving a problem? Obviously, it does not but it gets the problem out of their office and that is all that matters. The refusal to do anything makes a mockery of the army. The Georgia National Guard IG office was then given the option(?) of not doing anything about an issue they had looked at. What is the point of kicking the issue back to them when they refused to do anything in the first place? These reeks of an office not wanting to be involved in a matter, if they do not want to be involved perhaps, they need to let other people work in the IG office. What are the odds the Georgia National Guard IG office will suddenly decide to do the legal thing? They did send me a letter. I am sure it was to announce that they were pressing charges against randall simmons for telling lies to a senator twice and covering up crimes.

GEORGIA DEPARTMENT OF DEFENSE
OFFICE OF THE INSPECTOR GENERAL
POST OFFICE BOX 17965
ATLANTA, GEORGIA 30316-0965

REPLY TO
ATTENTION OF

October 14, 2011

Mr. Wesley Mahan
P.O. Box 125
Cohutta, GA 30710

Dear Mr. Mahan:

This letter is in response to your Inspector General Action Request addressed to this office on 8 September 2010 and the National Guard Bureau Inspector General's office on 9 September 2011. We conducted a thorough inquiry into your requests for assistance. Our inquiry determined that your Chain of Command has addressed the matters concerning forgeries, Standby board, travel voucher, and false statements.

In regards to your reduction in grade from E5/SGT to E4/SPC, it was found that you have missed a unit event, Yellow Ribbon, on 25-27 June 2010. This was supported by your own admission, and the command took appropriate Non-Judicial Punishment steps in accordance with the Army Regulation and policies.

As for the forgery on a travel voucher and EPS Board packet, these are legal matters which I will refer to the State Judge Advocate. However, if you feel that you were incorrectly paid in regards to this travel voucher, it is recommended to address the matter with the Defense Finance and Accounting Services **(DFAS)** who provides payment services to the United States Department of Defense. You would need to call or send your claims to the Out of Service Claims Department. Their address and phone numbers are DFAS-IN, Claims Branch, 8899 East 56th Street, Department 3300 (Attn: COR/Claims), Indianapolis, IN 46249-8654, Com: 317-510-6167, and Fax: 317-510-8654.

Lastly, matters regarding the 1SG Green's email comments have been addressed with the appropriate Chain of Command. We trust this information responds to your concerns, and this office will take no further actions.

Sincerely,

RICHARD H. SONG
MAJ, IG
Detailed Inspector General

FOR OFFICIAL USE ONLY. Dissemination is prohibited except as authorized by AR 20-1.

This document contains information EXEMPT FROM MANDATORY DISCLOSURE under the FOIA Exemptions 5, 6, & 7 Apply.

Where does one start with this thing of beauty? That letter is dated October 14th, 2011. It took them over a year to respond to a complaint about true crimes. Richard Song states that his letter is in response to my letter from 2010. If the NGB had not contacted them I doubt they would have responded at all. Yes, the 1-108th did address all those issues. They did so by threatening me, sitting on a transfer request, running me out of the military and violating my constitutional rights. The Georgia National Guard IG office knew this of course because they had stood by and let it happen. Also, there was no mention of randall simmons sending lies about me to the senator. How could this be? I mentioned that every time I wrote the Georgia National Guard just so they would not forget. Yet every time they were confronted with the truth, they either lie or threaten me, that is not exactly following the army core values. But hey it is whatever turns them on, right?

This proves to be typical of the military officer with an office job. He also displays the condescending attitude I have encountered from most members of the Georgia National Guard throughout this little game. For some reason he mentions the complaint I filed on September 8th, 2010, the complaint I never heard back from them on. I told them I felt like the 1-108th was trying to violate my rights and make a mockery of the justice system by refusing to grant me legal counsel and potentially have a hearing after I was out. Since they did have a hearing about me after I was out, and I had mentioned that as well.

Major Richard H song then states that after an inquiry he determined that my chain of command had already dealt with the issues! The reason I went to the IG in the first place was because of the way the 1-108th was dealing with the issues, by threatening me and defending crime. He mentions the yellow ribbon event I guess to rub salt in the wound but that comes off as stupid. It was not a unit event because when I went to the yellow ribbon event in august my unit was there, (suddenly it was an important event) we were not even supposed to go back to drill until July but at this point, he must defend the officers even if it means lying. If one wants to get right down to it, I did go to the event in June 25-27 anyway but i had been taken off the list at the desk. I hung around until it was determined i would not get paid and left, the briefings were useless anyway because they were all for couples as it was a non-military couples retreat. The bulk of the letter is about pay issues which was the thing I cared about the least.

When were brian green's comments ever dealt with? I was never told anything. The guy uses his rank and influence to kill my transfer because he doesn't like the fact that I reported crime and it's seen as acceptable? Who addressed this, brian green himself? That is the pattern unfortunately, deny, threaten, deny, lie and then keep on with the lies. Even if it means telling lies to elected officials.

No mention of randall simmons telling the senator lies, randall simmons defending crime or the hearing having took place after I was out either. Yet another example of the leadership culture protecting criminals because of their rank. How could they talk about randall simmons telling lies to a senator when that would destroy their entire narrative about me? Also, that would have opened investigations into high-ranking people in the Georgia National Guard which just can't happen. But why not? Is integrity not a core value of the army?

On the surface this should be simple, they received copies of the letters and emails from me and ostensibly the NGB as well, everything an investigator would need is right there. Why do they refuse to act? Is protecting an officer or three officers more important than preventing and punishing crime? Perhaps they thought i would just go away, tragically mistaken. Non pulsed I sent copies of the documents to the Georgia National Guard IG again and they responded. Different guy this time with a letter dated October 18th, 2011. Surely, they would not protect a criminal again.

STATE OF GEORGIA
GEORGIA DEPARTMENT OF DEFENSE
POST OFFICE BOX 1970
MARIETTA, GEORGIA 30061

October 18, 2011

Assistant Adjutant General-Operations

Mr. Wesley Mahan
P.O. Box 125
Cohutta, GA 30710

Dear Mr. Mahan:

I am in receipt of your recent emails and the copies of information you forwarded by mail to me.

I have reviewed each of the above and have conferred with our Judge Advocate General (JAG) office and with our Inspector General (IG) office regarding your case. Additionally, I have reviewed the documentation from the HQ, 48th IN BDE to include letters of response from Congressional Inquiries initiated by you.

Upon review of all the above it is my conclusion that the Georgia National Guard has appropriately responded to your allegations by the military chain-of-command system, and by investigation at each command level. Review by the State JAG has taken place, and by your own admission the IG reviewed your case and obviously found no grounds for action or intervention.

Predicated on all the foregoing and my own review I see no action required at this level. Your allegations have been found to not warrant further review.

Sincerely,

Donald W. Venn
Colonel (Retired)
Assistant Adjutant General-Operations

So if this letter is to be believed the Adjutant General Office of the Georgia National Guard has just said the following: Forgery is not a crime, telling lies to a senator is not a crime, threatening a whistle blower is not a crime and violating due process is not a crime. How can this be? I included the letters from randall simmons to Senator Chambliss and the email exchange between myself and randall simmons because to me it seemed an easy case. Everything the IG needed to open an investigation was sent to them. All they had to do was ask the officers why and punish where it would be called for in punishing an officer that punished a whistle blower.

But this was not the case, so to speak. They see me as a sergeant coming forward with letters and emails that show criminal activity in the Georgia National Guard. But to investigate this they would have had to punish some high-ranking people and that is not a popular thing to do in the any branch of the military. Honestly, I would have been surprised if the letters and emails were really sent to Donald Venn, why would anyone want to put him in that situation? The situation of defending criminals who happen to be officers. Just write a reply and forge his signature I mean they said forging documents and signatures are not crimes so what is the difference?

More disturbing is Don's admission that he "reviewed the documentation". So donald venn is telling me that he read the letters and emails and looked at the documents and found nothing wrong. Randall simmons lied to a senator about a person under his command with the intent to harm that person and donald venn sees nothing wrong? He states the guard has responded to everything. Is protecting criminals Venn's idea of nothing being wrong? What organization would say it is all right to fake financial and personal information on its staff, lie to elected officials asking about the fakes and threaten whistle blowers?

According to donald venn, he looked at everything I had sent in and whatever the 1-108th and 48th gave him as well. So, he looked at two cases of signature forgery, one case of misrepresentation, a threat to a whistle blower, two lies to a senator, an officer covering up crime in a letter to a senator, a due process violation and punishment that was illegal and found nothing wrong. Obviously, he thought I was an idiot when he wrote that letter. Either that or he is very corrupt. Or he just no respect for the law. Either way not an organization I want to be a part of anymore. Those were not the only cases of crimes I was just the only one that pointed out they were breaking the law in a way beyond complaining at the smoke pit.

He is correct in that the Inspector General office looked at the issues. They looked at the issues and told me they would not do anything because they did not want to punish individuals of high rank even though those individuals had broken the law. That was their quote. So yes, that did upset me. I mean if the IG office is not going to do its job why does it exist? At one point the IG office even said that brian green broke the law by threatening me but that they were not going to do anything. That is life and besides brian green was golfing buddies with one of the IG staff members. Punishment would have made golf so uncomfortable.

Venn compounds the problem by taking their side and protecting law breakers instead of prosecuting them. Keep in mind he has a state of Georgia government job. Would it not be easier to disband the Georgia National Guard IG office and just have the Georgia National Guard fall under state and local law enforcement? When not on federal orders the Georgia National Guard falls under OCGA so it can be done. That might change the way some people act. Now the IG office in Atlanta may like to write letters but they do not like to pick up the phone, but I do like to leave messages. One such message resulted in another letter this one being dated October 25th, 2011. Would they punish an officer? No.

GEORGIA DEPARTMENT OF DEFENSE
OFFICE OF THE INSPECTOR GENERAL
POST OFFICE BOX 17965
ATLANTA, GEORGIA 30316-0965

REPLY TO
ATTENTION OF

October 25, 2011

Mr. Wesley Mahan
P.O. Box 125
Cohutta, GA 30710

Dear Mr. Mahan:

This letter is in response to your call in request to look further into your matters. We have no new information that would cause us to reopen your case (Case # NGA12008). Please refer to our prior notification memorandum sent on October 14, 2011.

If you are not satisfied with our conclusions, you may seek redress through NGB-IG.

Sincerely,

RICHARD H. SONG
MAJ, IG
Detailed Inspector General

FOR OFFICIAL USE ONLY. Dissemination is prohibited except as authorized by AR 20-1

This document contains information EXEMPT FROM MANDATORY DISCLOSURE under the FOIA Exemptions 5, 6, & 7 Apply

What can one say to that? Still no mention of randall simmons telling lies to a senator but dick song did say he had reviewed everything. Yet he does not mention two huge things that I sent back to his office. The first obviously being copies of the letters randall simmons sent to the senator's office as well as the email traffic with the second being the financial and insurance paperwork I had attempted to present to the 1-108th while we were in Afghanistan. The paperwork that proved that the 1-108th was attempting to charge me with having an affair with a woman that they had created out of thin air.

The letters and email exchange were obviously ignored. This was unfortunate and short sighted on Song's part. I'm sure he did not see it like that, he no doubt thought i would just roll over and die. Throughout this process the attitude given off by the Georgia National Guard matched that. I just needed to understand that they were a criminal organization and just deal with it. The law did not apply to them.

So, to review I sent letters and emails from randall simmons in which he admitted to lying to a senator, protecting forgery and fraud, and holding a disciplinary hearing about me after I was out and it is no big deal? You know when I was talking to one of the IG guys on the phone, he said that these were crimes, then I told him the ranks of the people involved and he said never mind and that they would not do anything. The military does not like to punish officers or high ranking enlisted. There was also the totally unprofessional email from brian green but that is OK, I just did not understand that forgery and fraud were just the way the 1-108th worked. Apparently, they forgot to tell me, whoops!

I sent three FOIAs to the Georgia National Guard over the years, according to them they either never arrived or were filled out wrong. One of the FOIAs concerned those tags on the first letter. What were the enclosures? What was the content of the email from Teague? My reasons for those questions were because the first letter's rationale for forgery is a literal carbon copy of the rationale Teague used in Afghanistan when I confronted him about it. Teague literally said, "anyone in the chain of command...".

Teague even said that S1 could change what needed to be changed if they felt it needed to be changed. So, either clinton teague and randall simmons think exactly alike or teague had a strong influence on that first letter? How strong? Like strong to the point that he wrote it and just forged randall Simmons' signature on it perhaps? That would be ironic and would have given the 1-108th a feeling of moral superiority over me. I complained about forgery, and they responded with more forgery. So funny.

NGGA-108-CDR
SUBJECT: Response to Congressional Inquiry

3. For any questions or for more information please contact CPT Geoffrey T. Miller at 706-879-2916 or geoffrey.miller2@ga.ngb.army.mil.

RANDALL V. SIMMONS, JR.
LTC, IN, GAARNG
Commanding

Enclosures
Congressional Inquiry Packet
Email From SFC Teague

So, the congressional packet, some enclosures and an email from Teague were included with the letter. It bothered me that the individual in charge of the section forging documents and signatures, that would be teague, is that involved in the response to question being asked about criminal activities he supervised. The enclosure was the letter. Considering that the contents of the letter follow what teague said almost verbatim it would not be a surprise. But that is just a theory. A theory that randal simmons did not even think enough of the senator or crimes that happened under his command to investigate the crimes and write a response to the Senator's office himself. That would be very illegal but like I said it is just a theory that randall simmons did not actually write the first letter. The letters do not read like they were written by the same person at all.

They mention emails I sent, and I was able to print out a few before my time was up. After receiving the email from brian green i immediately sent two of my own out. The first was sent to rony mcstotts and carlos enriquez. This email was sent BEFORE the drill at which carlos gave me the papers saying I did not deserve jag support. It follows:

Would it be possible for a jag lawyer to be at drill? If I am going to have a false official statement thrown at me I would like to have legal representation. I would also like to discuss retaliatory punishment and forgery.

The second email was sent to rony mcstotts and joey recker, it said:

Why is a false statement being held over my head? That does not make it easy to trust anything I am told. Can you see what I mean?

Now first off yes everyone those emails were sent to outranked me but that did not matter anymore. It ceased to matter when the 1-108th put pen to paper and started filling out paperwork saying that I was in a relationship with a female that did not exist while in Afghanistan. At that point they were just supervisors in a very corrupt guard unit that I had to get out of. But the questions and points I raised were valid and ignored. They lost the psychological edge I knew they would lie so what did I have to lose? I did know to save as much information as possible. When you know it is a trap it is not really a trap.

There was never a jag lawyer at any drill in fact there was an attempt to write me up for contacting the jag office on my own. Just because I wish so badly, I had kept the copy they gave me. Also as seen in letters to senator Chambliss they did not let go of those false statements for quite some time. This created a situation they would only deal with using threats or more lies. If we had been active duty or still on active-duty orders that might have worked. But here, on drill status they had "rank" on me yes, but I had lots of time to look up laws and talk to people and was willing to go all the way. Even if it ended everything. Besides at some point it crossed over to personal and not about rank, nobody should ever be told that document fraud and signature forgery are ok. Then on top of that the organization committing the crimes threatens the witness and attempts to convince the witness that is ok to.

I also mentioned auto insurance paperwork and cashed checks, no reply of course then it is time to stir the pot. I mention being underpaid for the first 13 months, forgery, medical records being lost, forms being replaced and the lack of due process when threatened with punishment while we were in Afghanistan. Not real threats of punishment because the women did not exist however there was a lot of pressure to just sign a paper saying I had fucked a non-existent married woman.

I would have loved to see them follow through with the UCMJ threat, how would the judge have reacted when they presented a made-up female? Also, as previously mentioned carlos enriquez refused to look at documents that showed they had made the women up. By that I mean the 1-108th was pushing a name they just made up, never found out why. They would not say why they put a made-up person on paperwork and threatened to charge me with a crime that never took place.

I sent two emails regarding strong bonds. They all hit the same points that had already been mentioned. Forgery, false statements, multiple pay errors. I have yet to find a person who will say they would be comfortable at a job if the organization is comfortable with forging documents and signatures. If it is customary practice for documents and signatures to be forged how can a person exist in that environment safely? That was the bottom line in my mind I cannot speak for what the leaders of the 1-108th thought but for myself when I found out that they were OK with forgery and fraud and were willing to threaten people who report it the game was on.

Mcstotts responded by speaking of drills and threats. He pushed the lie that I was afraid of someone hurting me physically when if anything it would be the opposite. By that I mean if brian green had ever shown at a drill I was at I would have asked him to say it to my face. Him being the out of shape and insecure person he was he would have either made a remark meant to be rude or touched me. If he had touched me the entire dynamic would have changed instantly. It seemed like he was told to mention physical threats because mentioning the crimes would have been untenable. He was the only individual in the chain of command that responded. At that point, my impression was that the leadership of the 1-108th did not know what to do because I would not lie for them. They had gotten so used to breaking the law they did not know how to react when one person, just one person, would not bow to their criminal activities.

The 1-108th making up a lady to charge or threaten to charge me with adultery was crazy. I was informed of this change by brian green after I told joey recker that my bank had checks from me that were sent to her from Afghanistan made out to her birth name but signed by her married name. That helped make things clear to them that I had been played but instead of letting the matter go away they chose to change her name and threatened to court martial me for adultery. What is wrong with people in authority like that? They told me I could avoid all that if I just wrote a statement saying I had an affair with a married woman that did not exist. All this happened when I was at Torkham, which as has been mentioned was a small base on the border with no support services.

Despite many inquires I have never been told why those things took place beyond being told that they were going to get me. It was best to just put my head down and get through the rest of the deployment. Honestly, some people viewed this as me being broken. That did not make sense because I never signed the statement about the imaginary lady, I never even saw it. There was only a claim by the leadership that these forms existed at all. Joey Recker spoke to me and claimed he had one of the statements written by them, but he never allowed me to see it even though I was asked to sign it. I would not have signed it. I know that really does not make sense, how can I sign something they will not show me?

When we got back to Georgia I put in for a transfer, it was sat on of course so plan b. Someone in the 1-108th had decided to accuse me of having an affair with a woman that did not exist while we were in Afghanistan, would they lie to a senator's office about that? Yes. I also threw in my transfer and the 1-108th defending people that were forging documents and signatures, why not? When lying to the senator did not get me bow to them, they went into a punishment phase because I missed a couple's event that I was not required to be at. It was awkward. If I had just recanted my whistle blowing statements all the anger would have gone away.

On the one hand I had been promised that the leadership of the 1-108th was going to make me pay because I blew the whistle on them. But on the other hand, leadership knew that if they did not transfer me my contract would expire in a few months. They also still would not say why the forgery had taken place or why randall simmons lied to the senator about me. The only thing that interested them was to ram through some form of punishment or get me to recant my complaints. But I had already asked for a lawyer to be a drill and that would expose things. What could they do? So, what if captain enriquez decreed that a hearing and any resolution of the hearing would take place without a lawyer being present? Is that even legal?

Yes, technically in this case it is. I guess. Something to think about whenever a military recruiter contacts your kids. On the form he scratched through the parts about legal representation! This was after I specifically asked for a lawyer to be at drill, I wanted everything on the table. But even with this they were on the clock, they did not have much time to follow through on their threats. Now I do not know if it was legal for them to deny me a lawyer, I never researched that part of the Georgia Military Code. It is suspicious that given everything going on they bent over backwards to keep a lawyer from being there even to the point of writing me up for going to a jag lawyer.

So, it comes down to hearing on my last drill, prior to this drill the 1-108th had been contacted and asked why they were sitting on the transfer request and probably a few other uncomfortable questions. That resulted in an email from randall simmons and a voicemail from joey recker. Neither addressed the issues they just said they did not want me to talk to anyone in atlanta anymore. Such integrity. Now no lawyer is present, and I am sitting in a room with randall simmons and the entire chain of command and a few other officers for what reason I did not know. I was forced to represent myself. I got the impression this was some sort of entertainment.

It was kind of a sham, I attempted to bring up the threats and the forgery and was interrupted because randall simmons wanted to talk about me missing a drill that was not a drill. Verbal duals are like fights, after a few seconds you know someone has nothing but, in this case, he had rank and had shown a habit of abusing it and lying. Anytime I attempted to mention randall simmons telling lies to the senator one of his underlings in the room would interrupt me. I am sure that made them feel wonderful.

So, what was the point? There was no mention of brian green making threats after I complained of crimes, he had been given the weekend off! Randall goes on about missing drills, joey recker says something about pay, not pay errors, joey said they never mess pay up. It was a joke; it was supposed to make me feel bad I think or make me feel small. All I had to do was say I lied about the crimes the 1-108th allows people to commit. Details did not matter to randall only his rank and ability to cover up crime did.

However, there was one detail even he could not lie his way through. Whatever he decided had to approved by the brigade commander and I could appeal Randall's ruling to force that. There would be an actual hearing. But this was the Saturday of my last drill so what would they do?

Well, the Sunday of my last drill there was a ceremony to welcome the new commander of the 1-108th but there was no hearing and i had no chance to talk to colonel king. however, I did get an email a few days after the drill saying randall simmons decision to demote me had been approved, no hearing needed apparently. They sure showed me. My last pay stub showed no demotion, it was crazy. Also, no final paperwork yet.

It took a while, but I was able to goad a sergeant in the 1-108th to send me a copy of the completed paperwork from the hearing. He told me I had to sign it. A wonderful demonstration of the arrogance and criminality of the 1-108th or at least the people in charge at that time. The hearing was held months after I was out! I was not even invited! They really thought I would sign the paperwork. This was like a dream! Of course, the records NCO or the like was convinced that I had to sign it and honestly expected me to do so. Perspective: On September 11th, 2010, randall simmons started the demotion process, I was officially out of the Georgia National Guard on September 30th, 2010. The hearing was on February 3rd, 2011. I imagine the NCO is still waiting on me to sign the paperwork and return it to him.

So, colonel king or someone pretending to be him signed off on the hearing even though I was not there? Not only that but I had been out of the Georgia National Guard for months by then! Looking at this logically there are only two things that could have happened. The first is that his signature was forged, and he knew nothing of the hearing. The second possibility is that he knew what had happened and just went along with it. It is a little bothersome if john king chose option two. That would mean that john king is perfectly fine with someone being punished and not given a chance to defend themselves. The fact that I was out when he authorized my punishment is just icing on the cake of crime.

I suppose doing that also allowed john king to avoid answering a very awkward question. That being why did randall simmons change the girls name? The breakup happened late into the deployment, there is an unfortunate military tradition wherein commanders suddenly decide to punish as many people as possible towards the end of their deployment or tour. I am sure there is some professionally worded reason for that, but I just figured they were trying to make it look like they had been doing something besides signing off on things others had written and done.

That having been said for reasons unknown to me the 1-108th decided to go after me hard with the threat of an adultery charge. Luckily or sadly, it was easy to prove that I had been played however this was not the end of it. After it was made obvious that I had been lied to and taken for money the 1-108th changed the girls name and the circumstances. This might have melted my brain. Here I was on a small base on the Afghanistan and Pakistan border a base with no support services for service members and I had just been told my leadership intended to lie about me to court martial me. That is a good retention bonus.

By changing the circumstances, I mean I was accused of having an affair with a married women while in Afghanistan. As in the women was in Afghanistan with us. The 1-108th knew that was not true, they created a crime to accuse me of. That is crazy. Never shown proof, they never produced any shred of proof except for me being pressured to write a statement agreeing with what they were saying. Then it would all go away was their claim, wink, wink. All i had to do was write a statement saying I had committed a crime that I did not commit. The whole thing was crazy. It was either that or they had no plan and had expected me to lie so they could punish me.

So, on February third of 2011 Col john f king signed a form stating there had been a disciplinary hearing about me on that date that he had overseen. I was out of the guard after September 12th or 30th (point of view) 2010. Yes, there were paperwork issues after the twelfth but that was my last drill. So why would they violate my due process right by having a disciplinary hearing about me after I was out? That makes no sense unless there was a driving reason to do everything, they could punish me for blowing the whistle. I was already out, why do that? I would have loved to go, and I would not have showed up all alone. That was one of the reasons why I was not invited to the hearing. Was that legal?

Chapter 10: Build Me Up Buttercup

At this point I decided to write to the commanding general of the Georgia National Guard directly. Surely, he would do something. The response was a manila envelope with my letter having been opened, stapled to the envelope I had sent it in and stamped received by the adjutant general of the georgia department of defense. For some reason they photocopied the letter and sent that copy back with the original. Looking at the letter one can see that I was concise, and professional. You know long-term doing that to a letter from a person that is reporting crime might be the best thing to do.

I apologize for allowing myself to be forced to contact you regarding these issues but I have exhausted all the Georgia Guard's options. When I was in the 1-108th I discovered that some individuals were forging my signature. When I used the chain of command I was threatened. When I blew the whistle Lt. Col. Randall Simmons responded by writing a false statement as did Captain Carlos Enriquez. Lt. Col Simmons later wrote a second false statement in attempt to clarify what he said. I have spoken to the jag, the ig and Lt. Col. Simmons himself but nobody is going to do anything about the crimes and the cover up. The false statement Captain Enriquez wrote was used to demote me 19 days before I was honorably discharged. Lt. Col. Simmons has since admitted in an email exchange that he lied on the statements he wrote but that doesn't help me if the ig wont do anything. I have copies of everything I have mentioned and more. I am looking for what I've been looking for since the beginning, why was my signature forged and how many times did it happen? It would explain a lot of the pay and record issues I had when I was with the unit. I didn't ask anyone to lie to punish me because I reported crime but I've had to deal with it.

As you can see, I made no mention of any legal action. This was because I did not and do not want legal action, I want to know why those crimes happen and why the Georgia National Guard refused to do anything. It is more important to stop crime in the future then wanting payment for what happened to me in the past. They need some sunlight on the roaches. I got a response from William Spear and another letter which sadly I lost. The lost letter was beautiful, the officer took the side of other officers and said it was ok for them to lie about me and suppress evidence.

ATTN SFC Huffman

STATE OF GEORGIA
GEORGIA DEPARTMENT OF DEFENSE
1000 Halsey Avenue SE, BUILDING 447
MARIETTA, GEORGIA 30060
678-569-6001

January 27, 2012

Office of the Adjutant General

Mr. Wesley Mahan
P.O Box 125
Cohutta, Georgia 30710

Dear Mr. Mahan,

This letter is to inform you that we have received your letter here at the Office of the Adjutant General. After review it has been determined that there was a response to your congressional inquiry and we consider this matter to be closed. If you have any further questions you can contact us at (678) 569-6001.

Sincerely,

WILLIAM SPEAR, MSgt, GA ANG
Office Manager for the
Adjutant General's Office

As you can see Master Sergeant William Spear's reply is short and dishonest. The response to the first congressional inquiry was randall simmons lying to the senator and saying that forgery is not a crime, and anyone can do it. That is not legal, ethical, or acceptable. After the second inquiry randall simmons lied to the senator again and then give a very shady explanation of how my signature was only forged once, that was not true as we have seen. The lack of honesty and the ease at which the members of the Georgia National Guard lie to cover up crime was frightening to me at this point.

How hard is it to answer a question? I sent in two freedom of information act requests, and they were both ignored. At this point I was nearing graduation from college and as I was on the way to becoming a state employee I had to stop inquiring about the crimes. Ethical reasons of state employee vs state employee being the reason. I could always bring it back up later.

They would hear the questions again in a few years. Will Spear totally ignored everything and could only say that the congressional had been answered, yes it was answered with a lie and a cover up. Integrity. They thought I had given up on the truth. No but at that time it did not matter anymore I had other things to expend energy on. A few years later the idea to author a book came to me but I needed more information from the Georgia National Guard. So, I contacted the office of Senator Perdue in 2018. The inquiry asked four questions.

1. Who in my chain of command made the decision to lie to senator chambliss and myself? Why did they write that I was having an affair with a woman that did not exist?

2. Who made the decision to change the exgirlfriend's name? The change was made after it was proven that she got married while I was in Afghanistan, not before.

3. Why was my disciplinary hearing held in February of 2011 since I got out in September of 2010? They did not tell me when or where it was, and the 14th amendment was violated in this case. How can I have due process if the hearing is held in secret after I am out and I am not invited?

4. Why was I told by captain enriquez and lt col simmons that they would "get me" for reporting crimes they allowed and encouraged?

Four questions, four simple questions yet they could not be answered whenever I asked the Georgia National Guard. The fourth question was a little over the top, but it was true because they really said those things. Now one would think randall simmons would respectfully reply to this senator's office and not lie to him. Onto the reply.

STATE OF GEORGIA
GEORGIA DEPARTMENT OF DEFENSE
1000 HALSEY AVE SE, BUILDING 447
MARIETTA, GEORGIA 30060
(678) 569-6001

August 30, 2018

Office of The Adjutant General

Senator David Perdue
3280 Peachtree Road, NE, Suite 2640
Atlanta, GA 30305

Dear Senator Perdue:

This is in reply to your Mr. Wesley Mahan's request for assistance. Mr. Mahan has filed numerous requests for assistance since 2010. Each time, the Georgia Army National Guard has provided a thorough response to his inquiry. In the information provided to your office, it appears that Mr. Mahan is specifically alleging that he has submitted two Freedom of Information Act (FOIA) requests to the Georgia National Guard. He includes what appears to be one FOIA request that is signed, but undated and is not addressed to anyone. A search of available records has produced no FOIA request received from Mr. Mahan.

He then presents four questions he indicates were sent in with his FOIA requests. None of those questions constitutes a request for records under FOIA. Because Mr. Mahan is no longer a member of the Georgia National Guard, if he would like to request documents relating to his service, he should request those through the Army's Human Resource Command. More information is available to him at www.hrc.army.mil. The Georgia National Guard does not maintain records of Soldiers who are no longer active members of the organization.

Please feel free to contact me should you have any questions or concerns on this matter.

Respectfully,

JOACHIM P. FERRERO
Deputy Adjutant General

The very first sentence did not give me much hope, it just let me know I did not need to even bother reading the rest of the letter. But I did. So randall simmons did not respond but the guy that did respond, joachim p ferrero, refused to answer any of the questions. He also lies, I did not send Senator Perdue's office a foia request. That is not a good start. He then states that a search of foia records shows no foia request received from me, are these the same records wherein documents are regularly forged? This story also changes later. I wonder why there is no record of it. Those questions were sent into the Senator's office as an inquiry not as part of a FOIA. But it is easier for the Georgia National Guard to continue to lie rather than punish an officer.

The only thing unsigned that the senator's office sent to randall simmons was the place where my signature should have been on the paperwork showing they had a punishment hearing about me after I was out. It was unsigned because I was not there obviously. Everything else sent to Senator Perdue's office consisted of letters and emails from members of the Georgia Army National Guard. No FOIA request was sent into the senator's office by me at all. There was a FOIA mentioned but not one sent into the senator's office. So why would they lie to the senator and say that there was?

What reason or reasons could Joachim P. Ferrero have been given that would have made him lie to a United States Senator? Is there an inability of the Georgia National Guard to answer question from elected officials? Is there some hidden regulation within OCGA or Georgia Military Law that states officers can lie to elected officials, cover up crime and violate the law? Is there a reason or does the Georgia National Guard just put unethical people in charge? How hard is it for an officer to say why the officer lied about an enlisted member that reported crime? Or is it possible that someone that out ranked him ordered him to lie?

Of course, he had a reason for saying that because by lying and saying I sent the senators office a FOIA request he could then refuse to answer the questions the senator's office asked him. Well, played, illegal but well played. The agency ferrero gives out the info for does not handle affairs of the Georgia National Guard, I am sure he knew that. So, to review, none of the questions answered, the Senator's office lied to and given agency contact info that is useless. But he did say to contact him if there were any concerns.

So, I asked the office to contact him again which they did, and he responded by saying the letter was his definitive answer. I did not know you could refuse to answer questions from a senator's office. I mean if one watches cnnfoxmsnbcpbs talking heads and elected officials make it quite clear that telling lies to a senator or refusing to answer questions from one is not cool. So why do so many people in the Georgia National Guard get away with it? Is the Georgia National Guard beyond the reach of the law?

DAVID PERDUE
GEORGIA

United States Senate

September 11, 2018

Mr. Wesley S. Mahan
PO Box 439
Cohutta, Georgia 30710

Dear Mr. Mahan:

Thank you for contacting my office regarding your case. It has been a pleasure to work for you.

We have received a final response from the Georgia Department of Defense, and it is enclosed for your review. I hope this will be helpful in answering questions and concluding your case.

In the event you need additional information or assistance regarding this matter or any other federal issue, please do not hesitate to contact my office at (404) 865-0087. Thank you again for contacting my office and allowing me to assist you.

Kindest regards,

David Perdue
United States Senator

DP:DH

This also displayed the dishonesty level the Georgia National Guard would stoop to. The questions sent in by Senator Perdue's office were direct and not part of a FOIA so why were they not answered? Why has the Georgia National Guard continued to lie to protect an officer that lied to a senator about me and covered up forgery and fraud? Is it a case of massive corruption or massive laziness? Or is it a combination of both? Really the foia non answer was also a lie but not really a surprise. The Georgia National Guard will not punish an officer even when that officer allowed crime to take place and violated the whistle blower protection act.

At this point I wanted to see how many elected officials they would lie to or refuse to answer questions from. I went to the office of Tom Graves the representative to the United States House of Representatives in the district I dwell in. I gave them the documents and filled out a form. The questions were close copies of the ones Senator Perdue's office sent in with one about donald venn replacing the one about carlos enriquez and randall simmons. That being how could donald venn write what he wrote if he really saw all the documents, letters, and emails? The answer I received was confusing. Look at it. Tom Graves was polite enough to send me two letters of nothing.

PPROPRIATIONS
COMMITTEE ON
ERVICES AND
OVERNMENT
TEE ON DEFENSE
MMITTEE ON
N, HOUSING AND URBAN
AND RELATED AGENCIES

TOM GRAVES
14TH DISTRICT, GEORGIA

Congress of the United States
House of Representatives

WASHINGTON OFFICE:
2078 RAYBURN HOUSE OFFICE BUILDING
WASHINGTON, DC 20515
(202) 225-5211

DISTRICT OFFICES:
702 S. THORNTON AVENUE
DALTON, GA 30720
(706) 226-5320

600 EAST FIRST STREET, SUITE 301
ROME, GA 30161
(706) 290-1776

http://tomgraves.house.gov

October 31, 2018

Mr. Wesley S. Mahan
PO Box 439
Cohutta, GA 30710-0439

Dear Mr. Mahan,

Thank you for your recent inquiry concerning the situation involving the Georgia National Guard. I will be pleased to assist you with this matter. In order to proceed with your request, I must have the enclosed authorization form completed.

Please return the form to my office in Dalton, Georgia. Once the form is received, an official inquiry can be made on your behalf.

Sincerely,

Tom Graves

Tom Graves
Member of Congress

TG/vj
Enclosure

COMMITTEE ON APPROPRIATIONS

Chairman, Subcommittee on Financial Services and General Government

Subcommittee on Defense

Subcommittee on Transportation, Housing and Urban Development, and Related Agencies

TOM GRAVES
14th District, Georgia

Congress of the United States
House of Representatives

Washington Office
2078 Rayburn House Office Building
Washington, DC 20515
(202) 225-5211

District Offices
702 S. Thornton Avenue
Dalton, GA 30720
(706) 226-5320

600 East First Street, Suite 301
Rome, GA 30161
(706) 290-1776

http://tomgraves.house.gov

November 12, 2018

Mr. Wesley S. Mahan
PO Box 439
Cohutta, GA 30710-0439

Dear Mr. Mahan,

I have received the below response from the Georgia Department of Defense, Office of the Adjutant General. I have been notified of the following:

In reviewing this latest submission from Mr. Mahan, we see that he is continuing to ask questions that have been asked and answered in previous reviews. While he is disappointed with the outcomes he has received, it will not change. Colonel (Retired) Donald Venn prepared a response letter to Mr. Mahan on Oct. 18, 2011 which, attached here, clearly explains that Mr. Mahan's allegations have been investigated and do not warrant further review by our organization. Mr. Joe Ferrero, the Deputy Adjutant General, who prepared the response letter dated August 30, 2018, supports the response that Colonel Donald Venn submitted in 2011 in that no further action is required by our office to these same questions and allegations.

The above paragraph is a direct response from the Adjutant General. I regret that a more favorable response was not received. In light of the above response, your inquiry will be closed at my Dalton office and no further follow up will be warranted. Please call on my office again for any other Federal issues that may arise. Thank you.

Sincerely,

Tom Graves

Tom Graves
Member of Congress

TG/vj
Enclosure

Three for three. While it is true that the Georgia Nation Guard has been contacted many times about Randall Simmons telling lies to a united states senator, they have never actually answered the questions related to that. Now is refusing to answer a question an actual answer? Well in a way it is. At what point will someone in the Georgia National Guard admit that they do not apply to law to officers and let certain officers commit crimes without any punishment? Now if I am reading that reply correctly the Georgia National Guard has now officially endorsed forgery, fraud, and whistle blower retaliation.

Once again if donald venn really looked at everything and really decided there was no need for an investigation that would mean that he is saying, forging documents and signatures is ok and lying to a senator to destroy someone's career is also ok. How can this be? At no point has anyone form the Georgia National Guard said why randall simmons made up a female to threaten me with or why he lied to the senator twice about that. Also, they never said why the entire chain of command of the 1-108th did nothing when I reported forgery. What is going on in the command center?

After this I remembered that I had randall simmons email address or at least I did if it had stayed the same over the years. I wonder how he would react if I sent him some intentionally acidic emails asking him why he lied to a senator about me and defended forgery? Was the email address even good anymore? Only one way to find out. I sent him two short emails that were intentionally written sarcastically, I wanted a response, and one was received! Not from him but it is just as good. Yes, I was sardonic to him that was the point. The results of this were not what I wanted but better than I expected. I wonder what he thought when he got this.

I have enclosed copies of the letters and emails in which you lie to a senator, cover up crimes, admit lieing to a senator and admit to violating my due process rights. As you know the Georgia Guard IG and command staff responded to me over the years and refused to do anything about you lieing to the senator about me. This just will not do. So I'm asking you yet again.

1.Why did you lie to the senator about me?

2.Why did you allow people to forge documents and signatures?

3.Why did you violate my due process rights?

As you know there is more but lets just start with those questions.

The second

Why did you lie to the senator about me twice? Why did you defend people forging documents and signatures? Why did you violate my due process rights by holding a hearing after I was out of the guard and not even telling me when or where it was or allowing me to defend myself?

The third

I have not heard back from you about why you lied to the senator about me. Twice. You have also never said why you defended people forging documents and signatures or why you arranged a hearing about me after I was already out. I have all the letters and paperwork.

Those emails are not polite, but they ask some needed questions. I did not even know if this email address was good anymore, but I had nothing to lose by sending the emails. Would he reply or ignore me? Would he lie if he did reply or have someone else send a response? Well, I did not have to wait long Mr. Brian Bischoff replied for randall simmons. He did not answer any questions but gave me a few more to ask. Since randall simmons did not reply does that mean he now has personal integrity? Or is he just a coward? Now he was wishing he had stopped the criminal issues in the 1-108th. Too late now. If he even cared.

Mr. Mahan,

The emails you have sent to BG Simmons have been forwarded to me for response. Your case and the allegations you have continued to make have been reviewed at multiple levels of this organization on multiple occasions. The documents you provided in your latest email detail the due process you received during the process. At this point, your emails serve no purpose other

than to intentionally harass members of this organization. If you believe you were not provided appropriate due process in the matters that occurred nearly ten years ago, the proper recourse is to file a petition with the Army Board for Correction of Military Records (ABCMR). You can learn about the ABCMR here: https://arba.army.pentagon.mil/abcmr-overview.html. Should you continue to harass Brigadier General Simmons, he may have other remedies available to him through both the civilian civil and criminal systems.

R/

LTC Bischoff

BRIAN S. BISCHOFF

LTC, JA, GAARNG

Deputy State Judge Advocate

Ok one word, allegations, sadly it is not an allegation that randall simmons lied to a United States Senator it is a fact. Is it a crime if a victim of a crime is asking the agency tasked with investigating the crime to investigate the crime? Brian Bischoff is correct in saying that multiple levels of the Georgia National Guard claim to have investigated it. What he fails to say is that every single level either said they would not do anything because they did not want to punish an officer or that they just did not want to prosecute randall simmons personally. I say they claim to have investigated it because surely if they really investigated it, they would have done something not covered it up.

Covering it up is an action but it is a crime to use legal positions to cover up other crime is it not? Randall is special. So, every level looked the other way and refused to prosecute a criminal while letting a witness be threatened. Good stuff. Those can motivate a person to dig in. The reality of the Georgia National Guard doing nothing when shown proof of a crime is just too much. It cannot be true, can it?

Obviously when brian bischoff emailed me he had no idea that I have degrees in criminal justice and psychology. This had become a project for me. But he would not have cared anyway. It was all about defending the criminal who happened to be an officer and had also covered up crime.

His reply was also a little insulting. Does brian bischoff the jag lawyer think I am so stupid I do not know what I sent in? I do not know what happened to me and am iodic? Or is this what the Georgia National Guard teaches its jag lawyers to do when confronted with crime by high-ranking individuals in the Georgia National Guard to threaten the victim?

Of course, Brian Bischoff the Georgia National Guard jag lawyer mentions due process. Due process is not an officer writing two letters to a senator in which he lies about a whistle blower and covers up for criminals. Nor is due process an inspector general's office that refuses to do the job because they do not want to punish an officer. Due process could be defined as all evidence being on the table and everyone involved getting a chance to speak. But that would have caused an officer to get in trouble which just cannot happen.

This is not a game even though they see it as one, it is amazingly simple, either it is a crime to lie to an elected official and cover up crimes with the lie or it is not. They will not answer that question because it raises even more question about what goes on within the Georgia National Guard. Due process comes into that somewhere, but they ignore that to. It is easy to lie when the lie will protect you and being ordered to lie makes it all better. I was just following orders is a nice phrase. My understanding is that it is extremely popular at fort hood.

Where are the legal ethics? How does Brian Bischoff, a jag lawyer in the Georgia National Guard, become convinced that it is ok to threaten a victim of a crime? Is it because randall simmons is an officer? Is it because the law does not matter to the Georgia National Guard? What is the rationale behind the crimes and the cover ups?

He throws out the word harass in what I believe was an attempt to scare me. Yes, I have contacted the Georgia National Guard multiple times about randall simmons telling lies to a senator about me, covering up crime and allowing a whistle blower to be threatened and while they replied as we have seen they never actually answered the questions. So, the definition of harass is the victim of a crime reporting the crime? Oh, and randall used the lies to end my career I wonder how many other people he has done that to?

They also never punished anyone involved, except the whistle blower who was me. So yes, I would like to know why because if officers can lie to elected officials and a unit can forge documents and signatures how is anyone safe? In this jag lawyers mind was I just supposed to roll over and let crime continue?

That email was beautifully written. It was also very dishonest. So just so there is no confusion a jag lawyer for the Georgia National Guard is telling someone that an officer of the Georgia National Guard can lie to a senator about them and violate their due process rights and the Georgia National Guard will not consider it a crime. Also, if he read everything, I sent to randall simmons he is also saying forgery and fraud are not crimes as well. That is unless brian bischoff, a jag lawyer in the Georgia National Guard, is being dishonest and never actually looked at the letters, documents, and emails.

All that being said what remedy would randall simmons have? If Bischoff is to be believed randall simmons was going to go to a lawyer or a law enforcement agency and tell them what? That an enlisted member he lied to a senator about was asking why or would he lie to whoever he went to as well? What is the end game? To keep telling lies and throwing threats until I give up? I replied obviously.

US code and OCGA both cover false statements to elected officials. Randall Simmons admitted that he lied to Senator Chambliss about me on at least two occasions. I sent him the letters and emails concerning that. You reply with a veiled threat of legal action because I want to know why he lied to the senator about me and why the Georgia guard refused to do anything? I mean if he sent you the emails and documents I sent him then you would know that I'm telling the truth. For almost 8 years the georgia guard has been telling me that an officer can lie to an elected official about an enlisted member and it's ok. Is it?

Now I might be biased but that is one great reply to a threat of legal action. Especially since BB the Georgia Guard jag lawyer was not honest within the context of his threat to me. I foolishly though that Bischoff had integrity. What are the odds that he would reply with another threat? What are the odds he does not even spell my name right?

His reply:

Mr. Mayan,

Thank you for your response. Brigadier General Simmons and the Georgia National Guard have provided a response to your concerns on numerous occasions. There will be no further response to your inquiries; we consider your matter closed. Should you continue to harass Brigadier General Simmons or other employees of the Georgia National Guard or Georgia Department of Defense, there may be other remedies available to them through civilian civil and criminal systems.

If you believe you the victim of a crime or a civil wrong, you may hire an attorney to explore options that may be available to you. As previously mentioned, if you believe you were denied due process by the Georgia Army National Guard you may hire an attorney to explore your options of requesting relief from the Army Board for Correction of Military Records (ABCMR) -- more information on the ABCMR can be found here: https://arba.army.pentagon.mil/abcmr-overview.html.

Respectfully,

LTC Bischoff

BRIAN S. BISCHOFF

LTC JA GAARNG

Deputy State Judge Advocate

1000 Halsey Avenue, Bldg 447

Marietta, GA 30060

Now if brian bischoff is telling the truth and randall simmons really forwarded him the emails and documents then a question is raised. What is brian thinking? As he himself says this is not a new issue there a bit of history and paper trail. So why does he lie about that and issue a veiled threat of legal action? I should say theoretical legal action because how does the victim of a crime asking why the crime took place become a criminal? Especially when the workplace has been protecting the criminal. It was not like they did not know; the letters prove they did they just did not care.

That aside bischoff gives a cowardly reply. He does not even acknowledge any of the laws or code that randall simmons and other members of the Georgia National Guard have violated. Those violations were the reason I emailed randall simmons in the first place. Is this how he would argue a case in court? Just ignore evidence and change the subject every time?

Yes, the Georgia Army National Guard responded every time and every time they refused to address the issues or answer any questions. So, they consider the issue closed even though nothing was done about an officer of the Georgia Army National Guard protecting criminals and attempting to destroy the life of a whistle blower. Randall Simmons also used his lie to end my career. Brian Bischoff is protecting criminals. Is that part of the job for a jag lawyer in the Georgia National Guard?

There was no need for me to get a lawyer, there was only need for the Georgia National Guard to hold officers accountable to the law. Is a response all that is needed? Refusing to do something when a crime is reported multiple times becomes guilt by association. He does not even spell my name right. No lawyer was hired, the Georgia Army National Guard would either enforce the law or not enforce the law. It is a simple choice to make.

Also, sometime after sending those emails to me in which he threatened me brian bischoff apparently had his balls cut off. Every communication from or with him after that had him presenting himself as either weak or lazy. I wonder why? He did not even spell my name right the second time he threatened me!

The definition of due process of refusing to answer questions is bothersome as well. It really looks like brian bischoff thins that if he just threatens me, I will suddenly stop asking the georgia guard why randall simmons lied to a senator about me and protected criminals. Oh, how wrong he is. He does not even spell my name, right?! After the second email from Jag lawyer Brian Bischoff, I sent another FOIA into the Georgia National Guard. Here is the reply they sent back to me.

SGT Mahan,

No attachment came through. Our security probably flagged it for some reason.

The old form had social security numbers, but the new one does not. If you included your social that could be why it was removed from the email.

I attached our latest forms.

If you would like the Georgia National Guard Inspector General Office to look into an issue, please fill out the attached DA 1559 and continuation page and return to every email copied on this email.

Include any supporting documents (such as emails, texts, etc.) you have which will help us determine how to proceed.

Importantly, read through the privacy act information and select whether or not you consent with us speaking with others outside the IG channel to address your issue.

If you can't sign the 1559 (digitally or by printing and signing), then just return them filled out and say in the email you "couldn't sign it."

If you can't open the fillable version of the DA 1559, I have also attached a printable version which can be opened with any version of adobe.

The FOIA did not have to be on the DA 1559, it just needed to be professionally written and concise. I do not know why they wanted it on that form. It was just another lie or one of their games. But I played their game and sent the exact same papers to the National Guard Bureau in the Washington D. C. area, here is the reply from the NGB. Also, the georgia national guard IG already had all those documents. How or why would their security flag an inquiry into criminal acts committed by and protected by officers? What I sent the National Guard Bureau was ok. Also, I had not been SGT in years, I was out.

Good afternoon sir, your request was received in our office on August 27,

2019 and assigned case #J-19-0242, please see attached response

letter. If you have questions or concerns please email or call us at

844-573-2939 or email ngb.foia@mail.mil.... Thanks and have a great day

The NGB said the Georgia National Guard needed to handle that and they told me they sent the FOIA back to the Georgia National Guard and told them to handle it. They also gave me a phone number. I called the number, and it was answered by Brian Bischoff. He identified himself when he answered the phone and I held off for just a bit on letting him know who I was. I just said the NGB had told me to call that number about a FOIA, after he agreed then I told him who I was.

Shockingly he did not threaten me with criminal charges again, he did say that they would NOT answer the FOIA though. Thought he would have gotten the point by then. The law only applies when he wants it to what a prominent level of integrity Brian Bischoff the jag lawyer has. I did email him after the phone call, it is below. I did not mention how the Georgia National Guard IG lied and the National Guard IG told the truth. He obviously knew so what was the point?

I spoke to the NGB FOIA people last Wednesday and Thursday. They told me that the letters randal simmons wrote to the senator about me as well as the email exchange do fall under FOIA. A FOIA about those will be sent to you by me from this email address on Monday November 11th 2019. The letter that the Georgia National Guard sent to another senator from Georgia in 2018 I believe will be the subject of a second FOIA, in that letter the person responding refused to answer any of the question from the senator's office. The NGB was perplexed about that. The results of those two FOIAs will determine if there are anymore in addition to what the NGB is sending to your office.

R

wes mahan

I sent it to him twice, the second was a noticeably short email. Still no legal action from anyone in the Georgia Army National Guard. It is like Brian Bischoff was just trying to scare me by making a threat that he knew was empty. Why would a jag lawyer threaten the victim of a crime? That would be illegal. Furthermore, the level of dishonesty in that threat is not appropriate for a jag lawyer in the Georgia Army National Guard. Hey, it was a short email why should I say the same thing to him so quickly?

attached is the first foia request as well as copies of the letters and emails in case the copies sent to randall simmons were lost

His eventual reply, note the lack of threats. Didn't he also say the Georgia National Guard would "never" respond to my question of why randall simmons lied to a senator about me again? He keeps throwing randall simmon's rank in there, I do not care. His rank made it ok for him to have lied to elected officials and covered up crimes. In a way jag lawyer Brian Bischoff is illustrating the problem that caused all this. Officers in the United States military are not bound by law. They can do whatever they want. Rank had not mattered to me since I left, BB's constant use of it just perpetuated the belief that the law does not apply to officers. Ironically in our phone conversation brian bischoff had told me the foia would not be done. Something must have changed his mind. But staying realistic I was sure they would answer it without answering it.

Mr. Mahan,

Please see attached acknowledgment of your request under the FOIA.

r/

LTC Bischoff

Here is the attached reply dated november 12th 2019;

Mr. Wesley Mahan

Via Email to lgtdrm@gmail.com

Dear Mr. Mahan: This letter is to acknowledge receipt of your Freedom of Information Act (FOIA) request dated November 10, 2019, for "all documents mentioned in those two letters as well as any emails, letters or documents that went into those letters being written." The two letters to which you refer were included in your email and consist of two letters written by BG Randall Simmons in response to Congressional Inquiries. Your request was received by the federally-recognized Georgia Army National Guard on November 10, 2019 and assigned case # FA-20-0001.

If you have any questions regarding your request please contact the FOIA Office at 678.569.5644.

BRIAN S. BISCHOFF Lieutenant Colonel, GA ARNG Deputy State Judge Advocate

Brian Bischoff gives an interesting reply. He intentionally or forgetful refuses to acknowledge the simple fact that randal simmons broke the law by lying to the senator. How can this be? How or why would a jag lawyer for the Georgia National Guard ignore the fact that one of his fellow officers intentionally lied to a Senator and had another officer lie to a different Senator just to cover up some crimes? Why would a jag officer ignore the crimes of the other officers in the Georgia National Guard? I wonder what he would do if a lawyer contacted him with copies of the documents jag lawyer brian bischoff got from randall simmons and asked him why the Georgia National Guard has consistently protected the officer's ability to lie to elected officials and protect criminals? Not a lawsuit or demand for money but simple questions that the Georgia National Guard has so far refused to answer? Someday.

What happened to brian bischoff's bluster? During our phone conversation he would not even acknowledge the emails only calling them a previous communication. It was like he wanted to wish those emails away or at that the very least deny that he ever sent them. But he could not because he did send them, I suppose he allowed himself a moral victory by telling me that the Georgia National Guard would not answer the FOIAs. Why put that in writing? But the letter regarding that FOIA was not a letdown. I do have the letters and emails, so the Georgia National Guard was only incriminating itself at this point.

If brian was going to threaten me with legal action twice then he should have brought it, yes, he was being a coward when he made the threats but that is beside the point. Apparently, he realized that he did not live on the internet, and he was defending a man that lied to a Senator. Or he was just a coward. Or he is just a dishonest jag lawyer. I mean if one is going to threaten someone with legal action it would be nice if that person had the spine to follow through. I cannot stand a cowardly bully.

How would that go through?

Brian Bischoff: Your honor we wish to press criminal charges on wesley mahan

Judge: Why?

Brian Bischoff: An officer of the georgia guard lied to an elected official about him twice, used the lie to push him out of the guard, protected people forging documents and signatures and allowed leadership of the 1-108th to threaten Mr. Mahan when he blew the whistle on crime. Mr mahan wants to know why so we wish to press criminal charges on him.

Judge: You wish to press criminal charges on him for reporting crimes?

Brain Bischoff: Yes

Judge:...............

Now by Febuary 9th 2020 I had not heard back from the Georgia Army National Guard about that FOIA, I had to ask them what was going on. This email was sent on 2-9-2020 to brian bischoff and randall simmons.

Mr. Bischoff;

I have not received the FOIA or information concerning the FOIA. This is disconcerting. Randall Simmons wrote the two letters to Senator Chambliss that the FOIA requests information on. Randall Simmons is still a member of the Georgia Army National Guard. Has Randall Simmons forgotten what went into him telling the Senator, in those two letters, that I was involved with a married female that didn't exist? I attempted to show the chain of command paperwork that proved they were accusing me of a lie while we were in Afghanistan but that was denied, the first letter Randall Simmons wrote mentions that.

I'm sure the delay in the FOIA response is just a paperwork snafu and I look forward to receiving the FOIA and finding out what went into the two letters Randall Simmons wrote to the Senator about me.

R

Mahan

I come at Bischoff like that via email and he does nothing. He was full of shit when he told me that I could face criminal charges for reporting crime. When Brian Bischoff the JAG lawyer said that the Georgia National Guard would no longer respond to my inquiries, he guaranteed that the inquiries would not stop. It is a criminal matter; officers are accountable to the law in the same way other members of the military are. I imagine reading that burnt brian up, he would get his smile back later perhaps. He would "reply" to the foia but that did not matter. The FOIA was just a warmup. So, after calling the cid office at fort stewart and asking if anything in this was within their scope, which it was not, I took their advice and contacted a state representative. So, I contacted Chuck Payne, my district's representative in the georgia state house of representatives. As you can see from the email, I was concise. What would randall simmons say? Would he say anything or order someone to reply for him again? This is what I sent to Chuck Payne.

I am sorry it took me so long to get back to you, if the scanner worked that attachment should be two letters from randall simmons to Senator Chambills and an email exchange between myself and randall simmons. There is a large amount of information in those but the only thing that concerns me is why did randall simmons lie to Senator Chambliss with respect to telling him that i had to be instructed by the chain of command to stop talking to "ivy ferguson".

He based this on me being dumped while in Afghanistan, while I was deployed my bank sent me a fraud alert stating that checks I had written out to my exgirlfriend had been signed by someone else. The girls last name had been REDACTED but she signed the back of the checks with her last name being REDACTED. One phone call confirmed that she had gotten married and not told me because she didnt want a stop payment put on the checks. I called her one more time and she told me it was over and that was it. That should have been it.

I have contacted the guard many times over the years asking about the letters, emails, threats, and document forgery that took place in that time period and they either refuse to answer or give me a couched threat. On the advice of an Army CID officer, I have decided to simplify my questions and contact a state level NOT federal level elected representative.

All I want to know is why randall simmons lied to Senator Chambliss about me thus destroying my character and why he did nothing about forgery going on under his command and in fact defended it. There was a disciplinary hearing held about me after I was out but that is being investigated by an agency, I think it is at least. So, I just would like to know why randall simmons lied to the Senator about me twice in order to destroy my character and why the georgia guard has consistently refused to do anything about it.

These are bits and pieces of the emails to Mr. Payne.

Thank you for responding. This issue is something I've been working on since 2011. I have contacted the georgia guard multiple times, I have received many replies however they have all refused to answer the questions, lied or threatened me.

The questions is why did randall Simmons lie to senator Chambliss about me in writing twice? I have the letters,emails and documents. This is not a money grab, I want to know why he did that and why the guard refused to do anything. I'm at work but I should be able to pull copies of the letters and emails that started it and get them to you. I have no desire to get money from anyone I just want to know why the guard protected him. I'll get the documents to you asap.

Thanks

Wes mahan

A second one after a misspick on the send button.

Sir,

These emails are an example of how the guard has been dodging. They say that an officer can lie to a senator about me, telling the senator that I was committing a crime, which randall Simmons knew not to be true, and that I have no recourse. That an officer can lie to an elected official or anyone about an enlisted person and there is no recourse. All I want to know is why he lies about me and why the guard has done nothing.

Please and thanks

Wes. Mahan

Those are very respectful and direct. Surely the Georgia National Guard would respect an inquiry from an elected official. Well, it would be a first. I imagine it would be hard to explain why you lied about a service member to a senator twice and told the senator that forgery was not a crime, and anyone could do it. Then to compound that randall simmons had other people lie for him, the shame level was pretty low. Their reply dated 20 June 2019:

DEPARTMENT OF THE ARMY
GEORGIA ARMY NATIONAL GUARD
JOINT FORCE HEADQUARTERS
1000 HALSEY AVENUE, BUILDING 447
MARIETTA GEORGIA 30060-4277

NGGA-AGD-AR

20 June 2019

MEMORANDUM FOR Georgia Army National Guard Soldiers

SUBJECT Personally Identifiable Information (PII) Breach

1 On March 29th, 2019, the Georgia Army National Guard (GAARNG) lost control of electronic records containing the personally identifying information (PII) of a large number of Soldiers in the GAARNG from across all commands. The specific information compromised includes Soldiers' name, Social Security number, MOS, duty status (M-day, AGR, or technician) and unit information (unit name, UIC, and zip code)

2 The breach was reported to National Guard Bureau and subsequently to the Department of Defense Privacy Office. As this incident involves Soldiers of the GAARNG, I have initiated an investigation. And while initial indications are that there is a low probability that the compromised information will be used for an unlawful purpose, it is impossible to say with certainty that this may or may not occur

3 Considering the possibility of misuse, I believe you should make every effort to protect yourself against the potential that someone might use your information to commit identity theft. Please visit the Georgia Consumer Protection Unit website (http://www.consumer.ga.gov/consumer-topics/id-theft-military-personnel) and the Federal Trade Commission (FTC) website (http://www.consumer.ftc.gov/topics/privacy-identity). The FTC recommends immediately placing an initial fraud alert in your credit file if you believe that your information may be used for an unlawful purpose. The fraud alert duration is 90 days. During that time, creditors are required to contact you before changes are made to your account and before new credit cards or other accounts can be opened in your name. The site also provides other information that can be used now or in the future if identity problems should develop

4 The GAARNG takes this compromise very seriously, and we are reviewing our current policies and procedures to improve our processes and guard against future occurrences. In the meantime, you may contact your chain of command or the Office of the State Judge Advocate if you believe that you have been negatively impacted by the PII breach

5 The POC for this memorandum is the Office of the State Judge Advocate at (678) 569-5644 or ng.ga.gaarng.mailbox.ngga-jfhq-legal@mail.mil

RANDALL V. SIMMONS, JR
Brigadier General, GAARNG
Commanding General

Now keep in mind that was a reply to an elected official asking questions about crimes that randall simmons had committed and allowed. Randall Simmons responded by sending a letter that was obviously sent out to soldiers enmasse and did not even acknowledge the elected official or the crimes that had been committed. How disrespectful is that to the elected official and to the law itself? I do not know, obviously, what randall simmons was thinking when he sent that letter but how disrespectful can he get? By refusing to answer questions and sending a form letter as a reply he did answer. Randall simmons obviously does not care about the law and has realized that the Georgia National Guard and the United States Army would let him violate the law as much as he wants. What is there left to say?

A few more things at least as questions beg for answers. What does this say about the personal integrity of randall simmons who i have been told is a general in the Georgia National Guard? Randall simmons takes personal integrity so seriously that he lied to a senator about a whistle blower twice in order to protect criminals. That is not nice at all. Randall simmons views personal integrity as so important that he refused to respond to multiple inquiries from elected officials and when he did respond he used a form letter that was not even about the inquiry. That does come off as disrespectful does it not? Randall simmons treasures his personal integrity so much that he ordered a lawyer to threaten a whistle blower. What a quality leader the Georgia National Guard and the US Army have in randall simmons.

How to explain that reply. I would imagine there is a clause that states the Georgia National Guard had to answer so they did by refusing to answer but sending a letter out to me that had nothing to do with the questions asked. Very disrespectful and tragic in that the Georgia National Guard has chosen to endorse crime. What lesson is there from this? That rule of law does not apply to an officer, so does anything? How little must randall simmons think of the rule of law refuse to answer questions from an elected official about crimes that randall committed and criminals he protected? Fort Hood.

Chapter 12: You Will Know My Name

There were still the loose ends of the FOIAs to wrap up. The first two FOIAS sent to the Georgia National Guard were ignored or as Mr. Ferroro put it there was no record of them. So, for the third one I decide to track everything. It was very illuminating. But first the FOIA guidelines from the FOIA website at https://www.nationalguard.mil/Resources/FOIA/.

Essential Elements of a FOIA Request

Cite that you are requesting records under the FOIA

Provide your full name and contact information, to include e-mail address to expedite communications

Provide a reasonable description of the record(s) requested as well as the office(s) you believe are most likely to have the records sought. A listing of the offices serviced by this requestor service center can be found here.

Provide a statement of your fee category and indicate your willingness to pay applicable fees for processing. You may include a dollar amount you are willing to pay up to, or you may leave the amount you are willing to pay open ended.

That seems pretty simple, as in something that would be hard to mess up. As previously mentioned for some reason the first two FOIA requests vanished, and the National Guard Bureau had to tell the Georgia National Guard to respond to the third. That was very telling, a jag officer from the Georgia National Guard goes from threatening me with criminal charges to telling me the FOIA was being processed almost immediately. It was very sickening that the Georgia National Guard put the officer that had threatened me over reporting crimes in charge of the FOIA. This showed the Georgia National Guard's lack of respect for the law, I did not expect a response just another lie or threat or someone saying they do not remember. Brian Bischoff replied. Again, note the lack of threats.

Dear Mr. Mahan: This letter is in response to your Freedom of Information Act Request dated November 10, 2019. A search by the Office of the State Judge Advocate located two pages responsive to your request. Those two pages are records related directly to you, and are therefore being provided to you in accordance with the Privacy Act. No further records were located responsive to your request. The record will be send to you via DoD SAFE. Upon retrieval of the record, you will need the following password to access the record: GA-20-0001.

As the PA allows you the first 100 pages of your record free of charge and there are no fees for search, this record is being provided to you at no cost.

This concludes the processing of your request. If you have any further questions regarding your request please contact my office at (678.569.5644) or by e-mail

Shockingly the passcode did not work which led to another email saying that it did not work. After a prompt reply I was able to get the documents and was shockingly let down. As the email indicates the FOIA about how randall simmons wrote two letters to a United States Senator saying I was having an affair with a female that did not exist came back with 2 sheets of paper. The papers were the 4100 from fort mcclelan alabama that had me listed under the wrong MOS and the standby document on which my signature was forged. They had nothing to do with the letters randall simmons wrote to the senator. Another flip of the bird to the law by the Georgia National Guard. Also, there were no other records responsive to that request, how interesting.

So, what was the purpose of the Georgia National Guard sending me those two papers? Obviously, Brian Bischoff knew they did not have anything to do with the letters randall simmons wrote to the senator so why did he send them? Why would a jag lawyer for the Georgia National Guard intentionally send out paperwork in a FOIA answer that had nothing to do with the FOIA? Well, the answer is obvious as it has been since randall simmons told me via email he would never respond to me again. The Georgia National Guard has made the choice to defend criminals until the victims get frustrated and go away. In fact, one of the documents, they sent me was an example I had used when showing where my signature had been forged. That lack of integrity of the georgia national guard's part is sad.

If the FOIA had been executed as it should have been brian bischoff and randall simmons would only had two options. The first option would have been to produce whatever documents and emails went into the two letters to the senator and the hearing about me after I was out, that would end some careers and put some people behind bars. The second option would have been to refuse to answer based on fifth amendment assertion, the chose to not follow FOIA law and went with a third option. That option was to ignore FOIA and just send me to documents meant to upset me.

Now why would they do that, what would convince people that it is ok to lie or withhold information over a FOIA request? Regardless I decided to contact the governor's office. The governor of georgia is commander in chief of the Georgia National Guard so this should be simple it is just a few questions. Of course, I knew better buy why stop now? Also, I decided to send in some more FOIAs. The response to my first FOIA must have been a mistake because the two papers I was sent had nothing to do with the request. So, I just send another one in and email the jag lawyer again.

On July 8th of 2010 Randall Simmons wrote a letter to Sen. Chambliss about me. He told the Senator I had been involved with an "Ivy Ferguson". He repeated that claim in a second letter to the Senator which was dated may 15th 2011. In an email dated july 9th 2011 Randall Simmons admitted that there was no "ivy ferguson". The name "ivy ferguson" was thrown at me in Afghanistan as well as after we returned. This FOIA seeks all information that went into randall simmons telling the Senator that I was having an affair with a female that did not exist.

The reply to the FOIA:

Good Morning,

The FOIA you are referencing seems to be J-19-0242. We did refer this to GA ARNG. As mentioned back in late April:

The FOIA does require individuals provide a clear description of the record(s) they are seeking. The FOIA does require agencies respond to questions posed as a FOIA request nor provide interpretation of records produced. For records of the Georgia Army National Guard, you may make your request directly to their office at:

Georgia Army National Guard (FOIA)

1000 Halsey Avenue, Bldg 447

Suite W2145

Marietta, GA 30060

(678) 569-6490

In your request be sure to include a clear description of the record(s) sought, your contact information where a response may be sent, and indicate a willingness to pay fees for processing.

You do have the option to appeal, but please keep in mind there are timing requirements from when you received response from GA ARNG. Please see language below.

If you are not satisfied with this response you have the right to appeal to the Secretary of the Army within 90 days through this office, by writing to the Office of Information and Privacy (NGB/JA-OIP), 111 South George Mason Drive, AH2 Suite 319B, Arlington, VA 22204-1373, or by e-mail, NGB.FOIA@mail.mil. Your appeal must be postmarked or electronically transmitted within 90 days of the date of this response. Your appeal should include copies of your original request, this response and state the basis for your disagreement with the denial. If the denial includes the withholding of documents, then state the justification for release. Please note your appeal should address the information denied in this response and cannot be used to make a new request for new or additional information.

For any further assistance and to discuss any aspect of your request, you have the right to contact the Army FOIA Public Liaison Officer, Ms. Alecia Bolling, by e-mail at usarmy.belvoir.hqda-oaa-ahs.mbx.rmda-foia-public-liaison@mail.mil or by phone at (571) 515-0306.

Additionally, you may contact the Office of Government Information Services (OGIS) at the National Archives and Records Administration (NARA) to inquire about the FOIA mediation services they offer. The contact information for OGIS is as follows: NARA-OGIS, 8601 Adelphi Road-OGIS, College Park, MD 20740-6001, e-mail at ogis@nara.gov; telephone at (202) 741-5770; toll free at (877) 684-6448 or by facsimile at (202) 741-5769.

Please let us know if you have any questions.

V/R

MSgt Miller

One quote, "The FOIA does require agencies respond to questions posed as a FOIA request nor provide interpretation of records produced." What? How does that even make sense considering their non answer? Somebody forgot sentence structure. Appeal? Why can't the Georgia National Guard answer a question? There is nothing to appeal. It was with a touch of sadness that I noticed that brian bischoff did not reply. I wonder what msgt miller would say if asked why none of the information requests from the foia were answered. He did point out that the FOIA does require agencies to respond to question asked. So, I just sent it back to brian.

The other option is that the Georgia National Guard thinks so little of the law that they do not even proofread their legal replies. Nothing is worse than someone who does not proofread and must edit a book after it had been released. That is telling. Brian Bischoff went from threatening a whistle blower to refusing to answer a FOIA to just sticking his head in the sand. What happened to his bluster? Why wouldn't he answer now? Who tutored these people on sentence structure? If they are required to answer questions, then why don't they? And why refer an issue to the NGB when the NGB would have no documentation on it. Everything they needed was in Atlanta or in randall simmons files. Maybe he would to the second one:

This FOIA concerns a hearing held on September 11th 2010, a hearing held on February 3rd 2011 and the punishments and due processes connected to both hearings. The hearing on September 11th 2010 was staged and ran by randall simmons, the one in 2011 I was not allowed to attend.

The FOIA seeks all documents, emails, texts or other information concerning the hearings and punishments issued. Specifically:

1.Why was the appeal held after the I was separated

2.Why one of the punishments was illegal

3.Why I was denied the right to go to the appeal or take part in it in any way, I was emailed after the appeal had taken place and told that I had to sign off on it

I look forward to the completion of this FOIA

Obviously, there were doubts about the Georgia National Guard's integrity, so I went ahead with contacting brian kemp's office. This proved to be simple and complicated at the same time. As governor of Georgia brian kemp is the commander in chief of the Georgia National Guard. Surly brian kemp would see the wisdom of dealing with individuals in the Georgia National Guard who promote and protect criminals. I mean he is a republican who claims to be tough on corruption.

The office put me in touch with a guy who listened to me and asked me to send him copies and a letter explaining things. I did so and waited a few weeks and heard nothing which was not a shock. At this point I had been lied about, lied to, and threatened but I could sense frustration growing within the Georgia National Guard. They wanted me to go away, in their reality I should have gone away after I got out. That says a little something about the way they treat people. The ease at which they allowed crime and threats was not a good thing.

Some emails and voice mails later and I finally get back in touch with the guy. He told me he would send me his findings and told me that the Georgia National Guard refused to answer the questions and said I would have to get a lawyer to find anything out. Such integrity. He refused to put any of that in writing. So that is the reply from the office of brian kemp the governor of Georgia and commander in chief of the Georgia National Guard as well as a republican who claims to be tough on corruption?

What I got was the letter and documents I had sent to governor kemp with return to sender stamped on the envelope. The letter I sent to governor kemp had never been opened. Well now he has made it personal. I let it ride for a few weeks then call the office back. The representative stated that the office of brian kemp the governor of georgia and commander in chief of the Georgia National Guard cannot do anything about crime within the Georgia National Guard. That does not make sense at all and is a little terrifying.

So, I called the office of brian kemp governor of georgia and commander in chief of the Georgia National Guard again. Talked to an aide and after telling her what had taken place, she put me in touch with someone else. I contacted that person, sent her the documents, and waited. The wait begins to drag on a little. So, I send an email with the following questions:

1. Why did randall simmons lie to the Senator about me right after I blew the whistle on crime he knew was going on in the 1-108th?

2. Why has the Georgia National Guard refused to do anything? He used his lie to end my career.

No reply, not from the representative of governor of georgia and commander in chief of the Georgia National Guard brian kemp or the Georgia National Guard. It turned out that representative was on family leave and my file had gotten bumped to another office. Now I just feel foolish. So, I am told to email a different representative for governor kemp.

Sir;

This is an older issue but: In 2009 while a member of the 1-108th I discovered the personal section of the unit was forging documents and signatures on a regular basis. When we got back from Afghanistan I put in for transfer, the 1-108th sat on it and threatened me. I blew the whistle and went to the office of Senator Chambliss.

For reasons that have never been explained randall simmons defended the criminals and lied about me to the Senator. The lie was used to end my career. Except for a four and a half year period when I worked for the Georgia department of corrections I have stayed on this issue. I have been lied about a few more times, lied to and threatened by a jag lawyer.

I do not have a lawyer and have no intention of filing anything on the Georgia guard or the state of Georgia.

All I want to know is why randall simmons lied to the Senator about me twice, why the guard protected him while he was protecting criminals, why a jag lawyer, brian bischoff, threatened me after I reported crimes and if the guard is going to do anything.

I mean if I had known that officers in the Georgia guard could lie about enlisted members and protect criminals I never would have transferred into the Georgia guard. All I want is for the Georgia guard, preferably randall simmons, to tell me why he lied to the Senator and protected criminals.

All that sounds crazy but it all happened, I have a large collection of letters

Thanks

Wes mahan

I was not getting a confidant vibe from the officer of brian kemp the governor of georgia and commander in chief of the Georgia National Guard. So, I contact this aide and he tells me he will contact the Georgia National Guard and get some answers. I also email randall simmons just in case he forgot.

Are you going to answer the questions from Mr. Kemp's office or not? In case you forgot them:

Why did you lie to a senator about me?

Why did you allow threats to be thrown at me after I blew the whistle on crimes you let happen?

Why was I threatened by high ranking people in the guard after I blew the whistle on crimes?

He did not reply nor did the office of brian kemp governor of georgia and commander in chief of the Georgia National Guard contact me however a few days after emailing him I got two emails from brian bischoff, jag lawyer for the Georgia National Guard. They were sent within seconds of each other. I am not sure if they were supposed to demoralize me or not. The emails were written by a lawyer who needs to get a refund from wherever he went to college and law school.

Mr. Mahan:

The request below is requesting the same information you requested in Case FA-20-0001 and FA-20-0009. The agency has responded to both of those requests and provided you with information on how to appeal our response if you wished to do so. You had 90 days from the date of our response (May 13) to do so. The deadline for any appeal you wished to make was August 11. This matter is closed.

v/r

COL Bischoff

And counting these emails I see him mention three FOIAs. I did send in three but only heard back on two counting his next email. But at this point I imagine there was a lot of frustration and perhaps some fear floating around between randall simmons, who is a general and brian bischoff, who is a jag lawyer. At that time, they were both members of the Georgia National Guard which is an organization that protects criminals. If they were not such an organization this entire book would never have been written. JAG lawyer bischoff presents even more dishonesty as I could not appeal something I never received but it made feel good to type that. A response is not an answer, and he knew that.

SECONDS LATER...

Mr. Mahan,

Your request has been received by the agency and assigned activity number FA-21-0004. The agency will conduct a search for "all documents, emails, texts or other information concerning the hearings and punishments issued."

v/r

COL Bischoff

At this point it seems like they are just wasting time looking to find ways to answer without answering. The timing of this was suspect. Did bischoff really think that this reply would keep me from asking why randall simmons lied to a senator about me? Nothing in any of my interactions with Lt. Col. Brian Bischoff, a jag officer in the Georgia National Guard, suggested he had any integrity or personal courage. He started with threats of criminal charges against me for reporting crime but when we ended up talking on the phone, he said nothing about the threats. As I said before, where did his brass go?

He muttered some things about FOIAs in his jag voice but was clueless or being dishonest about the process. I had to quote the foia regulation at him because he tried to play dumb about what was covered. After he begrudgingly admitted I was right he said he would work on it then hung up on me! Is this learned behavior on his part? His utterance of closing the matter was a broken record at this point. How many times do they have to say they will never answer until I give up? What if I do not give up? What if every letter and email was just stacking to show this huge pattern of incompetence and dishonesty.

While this was going on I was still waiting to hear back from the office of brian kemp, governor of georgia and commander in chief of the Georgia National Guard. No reply, no emails, nothing at all. I emailed the guy again and waited. I am still waiting. So just to be clear brian kemp, the governor of georgia and commander in chief of the Georgia National Guard and a republican dedicated to law and order does not care about crime within the Georgia National Guard. That is good to know.

Office of the State Judge Advocate

December 14, 2020

Wesley Mahan

Dear Mr. Mahan:

This letter is in response to your Freedom of Information Act (FOIA) request dated October 15, 2020 to the federally-recognized Georgia Army National Guard for the following records related to two hearings allegedly having taken place on September 11, 2010 and February 3, 2011.

"[A]ll documents, emails, texts or other information concerning the hearings and punishments issued."

The federally-recognized Georgia Army National Guard received your request on October 13, 2020, and assigned it a local Activity Number FA-21-0005.

A search for responsive documents by knowledgeable staff in the Office of the Commanding General, Georgia Army National Guard as well as the Commander of the 1-108 CAV. No records responsive to your request were found. Records potentially responsive to your request would not be kept beyond six years in accordance with Army Records Retention Standards. The FOIA applies only to existing records, and there is no requirement to create a record in order to respond to a FOIA request.

If you are not satisfied with this response you have the right to appeal to the Secretary of the Army within 90 days through this office, by writing to the Office of Information and Privacy (NGB/JA-OIP), 111 South George Mason Drive, AH2 Suite 319B, Arlington, VA 22204-1373, or by e-mail, NGB.FOIA@mail.mil. Your appeal must be postmarked or electronically transmitted within 90 days of the date of this response. Your appeal should include copies of your original request, this response and state the basis for your disagreement with this response. Please note your appeal should address the information in this response and cannot be used to make a new request for

new or additional information.

For any further assistance and to discuss any aspect of your request, you have

the right to contact the Army FOIA Public Liaison Officer, Ms. Alecia Bolling, by e-mail at

usarmy.belvoir.hqda-oaa-ahs.mbx.rmda-foia-public-liaison@mail.mil or by phone at

(703) 428-6462.

2

Additionally, you may contact the Office of Government Information Services

(OGIS) at the National Archives and Records Administration (NARA) to inquire about

the FOIA mediation services they offer. The contact information for OGIS is as follows:

NARA-OGIS, 8601 Adelphi Road-OGIS, College Park, MD 20740-6001, e-mail at

ogis@nara.gov; telephone at (202) 741-5770; toll free at (877) 684-6448 or by facsimile

at (202) 741-5769.

This concludes the processing of your request. If you have any further questions

regarding your request please contact my office at (678.569.5644) or by e-mail

(brian.s.bischoff2.mil@mail.mil).

Sincerely,

BRIAN S. BISCHOFF

Colonel, Georgia National Guard

State Judge Advocate

Nobody had any records or knowledge? Not even randall simmons? So, track the logic, no documents are there, and nobody can remember or recall anything about it. Wow it is such a tragedy that the first FOIA sent in before 2017 just got lost or not lost since it was claimed they received it but never did anything with it. "Standards" would also say that forgery,fraud,threats etc. are crimes but it is selective as when they matter. So convenient for them. Randall Simmons cannot answer questions again. Maybe he forgot about the governor's office, I had better remind him. He might have forgotten.

Mr. Simmons;

Are you going to answer the questions from Governor Kemp's office or not?

Why did you lie to a senator about me right after I blew the whistle on crime in the 1-108th?

Why did the Georgia National Guard refuse to do anything about the lies and punishment hearing that took place after I was out?

That is all I want to know.

R

Mahan

Well to the surprise of nobody there was never a response sent back to me. So, the takeaways start with a FOIA request not meaning anything. Blatant disrespect to the process combined with dishonesty has shown that. I mean randall cannot remember anything about the letters or the hearing? How many letters to elected officials did he write and punishment hearings when the service member was not present did, he take part in? How many enlisted members of the Georgia National Guard did he lie about? Is the law just a joke to him, a general in the Georgia National Guard?

Another takeaway, two takeaways are that inquiries from elected officials mean nothing to the Georgia National Guard and that the law does not apply to any officer in the georgia national guard. How else can all this be explained? That the Georgia National Guard is not accountable to anyone is the logical conclusion. Officers can lie, cheat, allow crimes to take place and even threaten whistle blowers and not have to worry about the consequences of being a criminal. That is just the way it is. If anyone complains or reports the crimes that person becomes a target. The problem is not crime but people that report crime because that puts the buddy system in danger.

Is that a false theory? Let us look at the facts. Someone in the 1-108th was forging documents and signatures. I discovered this and used the chain of command to report it. The result was the threat from brian green. Then my transfer request suddenly stalled and eventually died. I did not back down. Then I am told that I refused to go to a three-day drill when there was no three-day drill, carlos enriquez told me I had to look bad for reporting crime. I did not back down if I had there would have been no point in staying in anyway.

Then randall simmons ups the ante by telling Senator Saxby Chambliss that I was committing crimes while deployed to Afghanistan. Randall simmons also said forgery was not a crime and anyone in the chain of command could do it. Those were his words on the first letter. He then illegally punished me as a "non-punishment" for reporting crime and held the final hearing after I was out. Due process?

This was followed up by years of lies, denials and a few threats. Why wouldn't they just answer the questions? It was like they would do any amount of mental gymnastics to avoid the questions even if it meant telling more lies to more elected officials. Then just refuse to answer, say it is a foia when it is not. Later when all else fails just threaten the victim and when the victim quotes law just threaten the victim some more. Then play rope a dope with some more foia requests.

The word "refer" is used a lot, to many times really. It became an answer without an answer. If the Georgia National Guard was faced with a tough choice say a FOIA or inquiry from an elected official that would incriminate high ranking members of the Georgia National Guard they would "refer" the matter. It might be to another office in atlanta or it might be to the washington dc area, but the answer was to refer the issue away. That was one of the main reasons I contacted the office of brian kemp.

What took place after that was a little troubling honestly. The governor's office was contacted multiple times. During this time, I noticed that randall simmons or someone that looked just like him and had the same last name was pictured with brian kemp in many of his photo ops. Now if that was really randall simmons would Brian Kemp, the Governor of Georgia and commander in chief of the Georgia National Guard, intentionally refuse to punish randall simmons? That could make evenings at the country club awkward.

But maybe I'm wrong. Instead of brian kemp, Governor of Georgia and commander in chief of the Georgia National Guard, being corrupt maybe he is just clueless. What if the aides to brian kemp are keeping things from him? What if he has no idea about the criminal culture of the Georgia National Guard? Surely, he would do something if he knew would he not? Or he could just tell his staff to ignore it.

But in a way isn't that perfect? What better way for them to prove the point I had been contacting them about? The Georgia National Guard had been sent a FOIA by myself and the NGB that was very precise and chose to ignore it and send two documents to me that were meant to be an insult. Instead, they proved my point. For some reason there are people within the Georgia National Guard that do not obey the law and protect each other so the criminal actions can continue as the Governor orders people to look the other way. That is one hell a recruitment bonus.

Chapter 13: Last Ride of the Day

I was given advice to wait for a Democrat Senator, so I did. However, the office of Senator Ossoff could never get a reply from randall simmons or the Georgia National Guard. They contacted the Georgia national guard and received no reply. They did receive a reply from the National Guard Bureau in Washington DC., here it is:

NATIONAL GUARD BUREAU
111 SOUTH GEORGE MASON DRIVE, AH2
ARLINGTON, VA 22204-1373

May 26, 2021

Office of Legislative Liaison

The Honorable Jon Ossoff
United States Senate
Attn: Teresa Dukes
3280 Peachtree Road N.E. Suite 2640
Atlanta, GA 3035

Dear Senator Ossoff:

This is in response to your inquiry on behalf of Mr. Wesley Mahan regarding his complaint. Because this issue has been adjudicated by both the State Adjutant General and the National Guard's Inspector General, Mr. Mahan's only course of action is to appeal to the Army's Board for Correction of Military Records (ABCMR). This Board is the highest level of administrative appeal within the Department of Defense.

The ABCMR has the statutory authority to change a military record, to correct an injustice or error. As a courtesy, enclosed is a copy of Department of Defense Form 149, *Application for Correction of Military Records*, with instructions for filing an appeal. However, this process can be done online at: http://arba.army.pentagon.mil. Mr. Mahen should provide compelling evidence or reasoning to substantiate his appeal. In addition, due to a heavy case docket, the appeal process can require up to 18 months to complete.

In addition, we are also forwarding the inquiry to the Georgia National Guard. We commend Mr. Mahan for his service and trust you will find this information helpful.

Respectfully,

D. Warren

Donna Warren
Chief, Congressional Inquiries
National Guard Bureau
Office of Legislative Liaison

Enclosure
DD Form 149

Copy Forwarded:
The Adjutant General, Georgia

To me that was very disrespectful and dishonest of the Georgia National Guard to ignore the inquiry and let the NGB handle it but sadly that was to be expected. In a way that says more about randall simmons and the officer core of the Georgia National Guard and United States Army more than anything else has. Out of everyone only two elected officials, one Republican and one Democrat, told randall simmons that he must reply. They were both disrespected. Integrity rules the officer core, right?

The National Guard Bureau did reply but with what? Yet again the military bends over backwards and lies by omission to defend the officer core. The hearing was the least of my concerns and that had been conveyed to the Georgia National Guard and the National Guard Bureau on previous occasions. So why would they ignore the real question and mention a hearing that took place after I was out of the Georgia Army National Guard? Maybe they need to be asked a few more times.

It turned out that randall simmons had left the Georgia National Guard and was now active duty at Fort Bliss Texas. It might not matter but that now meant that for the first time during all this randall simmons was under the authority of CID. The military traditionally does not to punish officers however so this would be a reach. But perhaps the CID office at fort bliss would do their job.

Randall Simmons being in fort bliss explained a few things as well about the governor's office refusing to answer any inquires. At the time when a voter is asking brian kemp, the governor of Georgia and commander in chief of the Georgia National Guard, about an officer in the Georgia National Guard that has committed crime and protected criminals why wouldn't brian kemp investigate the officer? Maybe brian kemp was more interested in protecting randall simmons from punishment. That would make afternoons at the social gatherings very awkward after all if kemp punished a criminal, he happened to be friendly with.

An investigation would have shown that randall simmons lied to a senator about a whistle blower. Then it would have shown that the Georgia National Guard had covered up for randall simmons instead of stopping the crimes. This would have put brian kemp, the governor of Georgia and commander in chief of the Georgia National Guard, in a bind. On one hand he sees absolute proof of corruption and crime in the Georgia National Guard then on the other the officer behind it all works with him or in is in his social circle. Choice.

Of course, that choice is known. The letter with the documents that would have incriminated randall simmons was returned to me unopened. So brian kemp, governor of Georgia and commander in chief of the Georgia National Guard, made a choice. Instead of investigating a criminal he let the criminal escape. Why would that be allowed to happen?

Regardless the escape changed a few things in terms of the search for answers. As said before randall simmons was now under the authority of active-duty army CID. So obviously I sent the CID detachment at Fort Bliss a letter and some of the incriminating documents, if they asked for the rest, I would send the rest. Otherwise, it would have been a huge envelope of papers and letters explaining the papers. During this time Senator Ossoff's office was just starting to ask about randal simmons criminal behavior. After getting back in touch with Senator Ossoff's office I asked them if they wanted the letter that brian kemp's office sent back to me unopened. They said yes, I am sure they found a beneficial use for it.

That only the National Guard Bureau responded has nothing to do with Senator Ossoff's office and everything with randall simmons lack of personal integrity. He lied then counted on people to lie for him. When that did not shake me, he had a jag lawyer threaten me and so on and so forth. Yet he was never able to scare me into recanting my statements reporting crime or forgetting that he lied about me and violated my constitutional rights. To speed things up I sent the IG at fort bliss a complaint about randall simmons refusing to answer the questions from elected officials.

I start getting calls from numbers registered to Fort Bliss, no letter, but phone calls. They were all from the Fort Bliss IG office. The calls left voice mails as most do. It ended up being six voice mails left in five days. A few days after that I got a message from Senator Ossoff's office. Those voice mails were the only reply I was going to get. This was before I received the letter from the National Guard Bureau in DC. I went back and listened to them. Whoever those people were that called me from Fort Bliss they are owed a refund from the education establishment. There was nothing they could do they said. Because the crimes took place before randall simmons got to Fort Bliss they were powerless, does that sound right? So, I contacted Fort Bliss via Public Affairs and the Fort Bliss "whistle blower" email. Never received a reply. It was a little disappointing since I did send the CID and IG offices at Fort Bliss copies of everything.

So, what is the takeaway? In every case except two randall simmons, the Georgia National Guard and fort bliss either lied to an elected official, ignored the elected official, or counted on the elected official doing nothing. In the two instances when the elected official pushed randall simmons to answer the question the results were disrespectful and possibly illegal. A form letter that has nothing to do with the questions and later just refusing to answer. That does not exactly scream integrity. Also, the IG and CID offices at Fort Bliss did nothing. That is right in line with Fort Hood. Tragic. This is what the military considers leadership.

One final note on good old randall simmons who is a general in the army and therefore innocent of anything ever. In August of 2021 when the taliban was taking Afghanistan randall simmons was contacted by an individual who had been an interpreter for him in Afghanistan. This person needed help getting another interpreter and some family members out. Randall had expressed his gratitude for that individual's bravery, randall simmons had said he would "help" if they needed anything. He did nothing. Exact quote, "I asked Gen. Simmons for help but he didn't even respond to text." Challenging work is hard when someone expects you to have integrity.

Chapter 14: Help me help you

Now what is the point of all this writing? This is not a series of complaints but a series of problems in search of solutions only for the Georgia National Guard to refuse to offer any solutions. In terms of the issues with the Georgia National Guard they were all caused by the Georgia National Guard having a chain of command totally enclosed within Georgia and the corruption that chain of command allows and encourages.

Now encourage is a strong word but look at the letters and emails. Look at verbal gymnastics of people who will not punish an officer or admit that the officer protected criminals. Look at multiple levels of the Georgia National Guard chain of command had no problem lying to elected officials to protect an officer and others. That kind of behavior obviously has ripple effects that cause even more dishonest and criminal behavior and could drive good people out of the Georgia National Guard. This behavior seemed automatic, there was little hesitation to lie or commit crimes. That is either learned behavior or the Georgia National Guard only recruits and promotes corrupt individuals. There was an obvious pattern that suggested it was quite common to threaten whistle blowers and to destroy their character and careers. Never any hesitation just automatic reactions.

Speaking to that what is the effect on members of the Georgia National Guard in general when they know that leaders at multiple state levels are dishonest? Do good people stay in? Do people that lack integrity stay in? Who stays in? That is tragically obvious at this point. How does this knowledge affect people who might have a desire to work in military law enforcement or law enforcement in general? Is it possible to do either of those knowing that the state of Georgia allows a criminal organization to flourish under the guise of the Georgia National Guard?

That could cause some problems. Yes, I called the Georgia National Guard a criminal organization. What else does one call an organization that allows forgery, fraud, threats to whistle blowers and constitutional violations to take place? Sounds like a criminal organization, yes? As was said before that could cause problems. These problems can cascade together to create a negative feedback loop that could result in criminals being put into positions of authority in the Georgia National Guard.

This could cause an institutional lack of integrity however integrity is a core value of the army so that would cause some problems. Making the situation worse is that without integrity there is no accountability which of course puts every member of the Georgia National Guard in danger. So, integrity and accountability must be put back into the Georgia National Guard. This is a simple fix. However other fixes will not be simple. The process has become broken when possession of a high rank becomes the ability to commit crimes at will. That does not bode well for the overall health of the Georgia National Guard.

Integrity and accountability could solve every problem within the Georgia National Guard. With integrity there is no case of an officer accusing an enlisted member of fucking a married women who does not exist, it just does not happen. With integrity an officer would never lie to a senator to protect people forging documents and signatures. With integrity an officer would not cover up crimes committed by several members of the chain of command. With accountability a service member would never be forced to wait an entire year for a pay error to get fixed, it would be fixed as soon as possible. Accountability would demand that people throwing away documents turned in by service members and replacing them with fakes would get punished not rewarded. Accountability would not let an officer get away with threatening a whistle blower.

One solution would be for the state of Georgia to enact a law, regulation or code that states that one must serve active-duty military for at least two years before serving in the Georgia National Guard. This would be for one simple purpose and that is to make sure everyone knows what it(military service) is really like. A bit simplistic but one can also hope that the integrity of active duty, which while not perfect is more than the Georgia National Guard has, would rub off on the service members. This would also tie into the next theory which would make the Georgia National Guard members accountable for their criminal actions.

As of this writing Georgia has the Georgia code of military justice which covers the Georgia National Guard. This needs to be scrapped, the national Georgia National Guard need to fall under normal OCGA when on state duty. Title 32 and Title 10 orders will remain the same. When I discovered that the 1-108th was forging documents and signatures on a regular basis what were my options? Local law enforcement told me they could not do anything because it was a Georgia National Guard issue. The Georgia National Guard did nothing just because they could. How can multiple law enforcement agencies refuse to do anything when crimes are reported to them? It is nice to have a state job.

Forgery and fraud are crimes under Georgia law, in this case when the Georgia National Guard refuses to investigate local law enforcement would have been obligated to take over. That would seem logical but it does not happen did. So, I propose a slight change to the Georgia Military Code and to Georgia law. The change would state that when the Georgia National Guard refuses to investigate crimes that take place when the guard members involved are not on federal orders local law enforcement will. In this case what would have happened if that was in place? I would have been threatened less for one thing.

Considering that one case of travel voucher fraud cost me 1000 dollars it goes into class two forgery. If Georgia law covered the Georgia National Guard, the way it should the local police would handle it when the guard unit decides to threaten the whistle blower and defend the criminals. Now that seems pretty simple. So what happens in this case when the 1-108th refuses to do anything? Well, the service member, me in this case, would report the crime to local law enforcement. So, I take the voucher to the local police department and report a crime. The voucher is obviously forged, and the forgery resulted in me being underpaid by at least 1000 dollars. Compounded with the massive amount of payment errors the guard let happen during my first year there and a pattern appears. That pattern could lead to an investigation of the 1-108th.

Now local law enforcement would have the option of looking at the crime and deciding on the appropriate action. Now this does not mean the worst punishment according to the law it means accountability. Knowing that a crime that the Georgia National Guard Inspector General office ignores will be picked up by local law enforcement could have a positive effect on the behavior of the administrative and command sections. Crime will no longer be covered up by the Inspector General's office it will be investigated if not by the guard, then by state or local law enforcement. Everyone wins. Except criminals within the chain of command.

Moving onto the lying to the senator about me, the real red meat. That of course is a violation of US Code 1001 and Georgia law and when the Georgia Nation Guard refused to do anything what would happen next under a new or rewritten legal statue? Well first the chain of command must have a chance and they did. I was told that it did not matter if randall simmons lied about me nor did it matter if he lied about the situation as well, he was an officer and could lie to an elected official if he wanted to. I missed that regulation because wherever it is or was i have never been able to find it. If anyone knows of the regulation that states an officer of the Georgia National Guard can lie to an elected official whenever he wants, please tell me which one it is.

So, in that case what would happen next? The unit failed to do anything, so the next step is the Georgia Guard Inspector General's office and as we have seen they refused to do anything. So local law enforcement? The FBI could or should be involved since he did lie to a senator twice and told the senator that forgery and fraud are not crimes and anyone in the chain of command can do them. If not the FBI but whatever federal agency investigates when someone intentionally lies to a senator to cover up crimes and destroy a whistle blowers' character and career. If that were a certainty to happen when the Georgia National Guard refuses to do anything a situation like that would never happen at all. At the very least the local police department should show up to arrest some guard members for forgery, fraud, and false statements. That would improve some attitudes and behavior patterns. Is the officer going to lie on an enlisted member when the officer knows the police officers might come talk to him?

That might seem over simplistic, but it tracks logically. A member of the Georgia National Guard discovers the unit is conducting criminal activity, reports it using the chain of command then gets threatened by the chain of command. That should never happen but there are times when it does. Options? The Georgia Guard IG and JAG refused to do their jobs so would it not make sense for local law enforcement to be involved? This would have two beneficial side effects.

The first being increased integrity and accountability within the Georgia National Guard. Would it not make the Georgia National Guard a better organization if people knew that just because someone had a high rank did not mean that they would get away with crimes? Many times, I saw members of the guard frustrated due to the obvious lack of accountability and integrity within the Georgia National Guard. This would be a strong step towards fixing that.

The second being that this would result in a savings for the state of Georgia. With law enforcement agencies now able to deal with local guard units when they break the law and threaten whistle blowers there would be less of a need for Inspector General and JAG offices stacked with people on state payroll tucked away in the Georgia National Guard command building. If one looks at them objectively, they are obviously a waste of money. An Inspector General office that refuses to investigate crime? A jag office that defends someone lying to elected officials? These people do not need to be working for any government they need to be in prison. At least with this they would not be getting paid by the state of Georgia to protect criminals who happen to wear the uniform of the military.

Or this could all be made obsolete by putting the Georgia National Guard under the authority of active-duty Army CID. When I contacted the CID office at Fort Stewart Georgia and told them everything, they wanted to get involved but could not. The 1-108th "works out" at Fort Stewart or did when i was with the unit so that was why I called them. There is no CID within the Georgia National Guard, it might be possible that has an adverse effect on things.

In this case when i receive a threat from the first sergeant i do not go to the Georgia National Guard IG office I go to an active-duty Army CID office. While the Georgia National Guard IG sergeant says he plays golf with brian green and I just must understand him active-duty Army CID would open an investigation. It would also help to have the organization responsible for investigating the Georgia National Guard to not be attached to the Georgia National Guard. It might require some additional training for the CID agents, but additional knowledge is good.

Brian Bischoff threatens a person reporting a real crime with criminal charges. He is or was one of the top jag lawyers in the Georgia National Guard. So, in that case what happens? Obviously, the Georgia Guard IG has shown itself to be worthless so where does one go to report a lawyer that is engaging in witness intimidation? That was a rhetorical question. The local district attorney's office or state bar office would easily be able to handle brian bischoff.

These are just theories, but something needs to be done. There should never be a case where a member of the Georgia National Guard finds the chain of command destroying the character of and lying about members who report crime. It all comes back to integrity. If the members of the Georgia National Guard that work in the Inspector General's office refuse to do their job then they should be replaced and charged.

Another way to prevent issues like this would be to just integrate the national guard IG and Jag offices with their active-duty counterparts. Instead of the state having one big office each active-duty station in the state would be set up to handle the big complaints from members of the Georgia national guard. This would also save the state of Georgia money since the Georgia Guard IG office in Atlanta could be shut down. Georgia would not have to pay people to sit in that office anymore. That would be a huge boon to the state's budget.

There is political failure and corruption here as well. In every case except two the elected officials contacted refused to do anything. This was not "someone hurt my feelings" this was "these state of Georgia employees are breaking the law". Not a question sadly it was and is true. So why...could these elected officials not step in when the state of Georgia department, the Georgia National Guard, did not? It's not even a case of wanting the Georgia National Guard to be forced to punish and officer that broke the law a few times it's just a quest for answers.

Given where the Georgia National Guard falls any of those elected officials had the ability to step in. But none of them did, with two exceptions, they all looked the other way and swept it under the rug. It would not have been difficult to do something as I provided them with the evidence every time. So why did they not? What reason could there be that these elected officials refused to do anything about state of Georgia employees breaking the law? That is a good question.

Another problem is that the United States military is still based on the old British class system. The Officer corps of the United States military is obviously structured after the British Crown and

Nobility factions when they had to "go to war". They were treated better than the lower folk and given more rights so to speak. It was the divine right of kings after all.

The enlisted class are equated to the simple folk, cannon fodder, we are there to die and not think. Now warfare has changed over the centuries but the attitude towards the commoner class has not changed in the military. An officer (divine right of kings) can do anything to an enlisted member (uneducated farmer) legal or not because the enlisted member will be to stupid to see what is going on or "does not matter". Just look at Fort Hood. Or talk to a current or former enlisted member that is honest.

Is that cold? Why else would jennifer jones be expected to accept a punishment for an alleged possible one-time sexual act when general fiscus could have multiple affairs and it was, ok? What else could explain that? Randall Simmons tells a United States Senator that I was a criminal when randall knew I was not and it is fine. I have a solution.

A simple reform, not a simple solution with its ripple effects would be to merge the officer and enlisted ranks. Now the service academy graduates and those with college degrees would still get a leg up but with a difference. Two and four-year degrees get those individuals boosted up the chain of command a little, but it is all the same chain. Now there would be some lingering remainders of the class system but those could be worked through. The names of the ranks would stay the same, some ranks would be merged, but everyone would be on the same track. The salute would go away except for ceremonies and other functions, and everyone is on the same scale.

But at the end of the day the underlying problem is that lack of integrity among members of the Georgia National Guard's command, Inspector General and JAG staff. Military leadership in general as well. Instead of seeing leadership as something to be held in high esteem they see it as a way to enrich themselves and to play games with the law and people under there sphere of authority. All they must do is drag their feet and people will get frustrated and leave maybe not but they should remember one thing. No matter how long they drag their feet and retreat into their safe spaces the truth will come out and there will be accountability. For every 400 service members that get lied about, threatened, or otherwise abused there only must be one that saves the papers and emails then authors a book about it. At least in theory. How many other people did randall simmons lie about and then use his rank to make the Georgia National Guard protect the lie?

Chapter 15: Isn't that the way it always is?

In addition to contacting brian kemp, governor of Georgia and commander in chief of the Georgia National Guard, I also contacted nathan deal when he was the governor of georgia. He replied but it was a bit of a non-reply.

STATE OF GEORGIA

OFFICE OF THE GOVERNOR

ATLANTA 30334-0900

Nathan Deal
GOVERNOR

February 24, 2012

Mr. Wesley Mahan
PO Box 125
Cohutta, GA 30710-0125

Dear Mr. Mahan:

Thank you for taking the time to write me concerning the Georgia National Guard. As a former member of our armed services, I appreciate the patriotism and commitment demonstrated by the men and women, past and present, who have served our state and our country in the National Guard.

Under state law, the operations of the Georgia National Guard are regulated by the Georgia Department of Defense (DOD). Therefore, in order to assist you in the best way possible, I am forwarding your letter to the Department of Defense for proper consideration and a direct response. You may also contact the DOD:

Georgia Department of Defense
935 East Confederate Avenue, SE
Atlanta, Georgia 30316-0965
678-569-6001

I regret that I cannot assist you further. I hope you understand the limitations of my authority in this situation, and I hope this information will be helpful to you.

Sincerely,

Nathan Deal

Nathan Deal

ND:ns
cc: Georgia Department of Defense

I am fairly certain this non reply resulted in one if the condescending letters I received from the Georgia National Guard during that time period. So, Nathan deal who at that time was governor of georgia and commander in chief of the Georgia National Guard did nothing and even put it in writing that he did nothing. Useless, dishonest, corrupt, lazy, afraid, which one of those words describes him? Deal is shown proof of crimes and does nothing but kick the crimes back down to the state of georgia department that had been committing them. What leadership. How hard is it to punish an officer when your office is given all the evidence that shows the officer committed and aided crimes?

Prior to contacting Nathan deal I had contacted the Inspector General Office of the state of georgia. Why they were not part of the Georgia National Guard they were, i thought, a law enforcement agency in Georgia. My mistaken though was that they would do something or refer me to an agency that would do something. Corruption protects corruption.

Office of the State Inspector General

Nathan Deal
Governor

Deron R. Hicks
State Inspector General

June 23, 2011

Mr. Wesley Mahan
P.O. Box 125
Cohutta, GA 30710

Dear Mr. Mahan:

This is to acknowledge receipt of two packs of documents you mailed to the Office of the State Inspector General within the past few weeks. You mailed these to our office after initially speaking with Deputy Inspector General Bill Donaldson concerning your allegations about your experiences with the Georgia National Guard.

We have reviewed the information you provided to our office. Please be advised that the Office of the State Inspector General can only take action on cases of fraud, waste, abuse or corruption that occur within the Executive Branch of Georgia government.

Your concerns do not fall within the jurisdiction of this office; therefore, we are unable to assist in the resolution of your concerns. Please be assured we will respond to any concerns that are within the legal authority of this office.

Sincerely,

Deborah Wallace
Deputy Inspector General

Enclosures

2 Martin Luther King Jr. Drive, SW • 1102 West Tower • Atlanta, Georgia 30334
Phone: 404.656.7924 • Toll-Free 866.435.7644 • Fax: 404.657.9716

Another brush off. Not only can they not do anything about it but they do not know anyone that can either. How is it possible that a law enforcement agency for the state of georgia has no idea what to do with a criminal complaint. Has anyone else had this problem when dealing with corruption in government agencies? I mean perhaps I see justice as potential rather than reality but that state of georgia agency could have done something. They just did not want to put in the effort.

Page 1 of

Subject	Re: Talk with CO
From	<wesley.mahan@us.army.mil>
Date	Thursday, June 24, 2010 20:03
To	"Mcstotts, Roney L SFC NG NG NGB" <roney.mcstotts@us.army.mil>

SFC Mcstotts:

I will not be at strong bonds, since I have been with the 108th I have had my signature forged on at least two occasions both of which were totally unneeded, S-1 lost, burnt or misplaced all my prior service medical records with no regard being shown for any privacy loss created, I was underpaid for the first 13 months I was with the unit despite constantly giving S-1 the needed paperwork to fix the issue, the 4100 I signed in alabama was scrapped and an unsigned one with my secondary instead of primary mos was sent to atlanta and the unit paid me back for going to the ig with a legit complaint by jumping me over my ex..

I was told all my rank would be taken away and called "not a real man" because my ex-girlfriend had gotten married an not told me. I attempted to tell the truth and show some papers that would have exonerated me but was told that was not going to be heard. I was then told that it didn't matter if things written about me on a counseling statement, which I was not given a complete copy of, were not true. I was made to understand that if the 108th wants to punish you that is all that matters. Soon after that I had a non-dated nco-er thrown in front of me that was full of garbage. I am filing a congressional complaint tomorrow it will include all the paperwork related to these things as well as my transfer request. I cannot trust leadership that has constantly let me down or placed the ability to punish me over the truth.

Sgt Wesley S Mahan

----- Original Message -----
From: "Mcstotts, Roney L SFC NG NG NGB" <roney.mcstotts@us.army.mil>
Date: Saturday, June 19, 2010 6:44
Subject: Re: Talk with CO
To: "Mahan, Wesley S SGT NG NG NGB" <wesley.mahan@us.army.mil>

> I willbe in Flordia next week until Thursday so I will not be in
> the loop on any thing . I haven't heard anything about the yellow
> ribion event yet.I haven't checked my AKO for a few days
>
> SFC Roney McStotts
> S-6 NCOIC
> 1-108TH CAV SQDN

Notice that I mention criminal acts not physical harm, they did not like that. It was direct and to the point and I made the mistake of thinking they would honor the law. Being there was an option, and I did not need the extra money and why would I want to be in the same location of criminals? I sent another email about that as well. Brian green gave me the excuse to say and do what needed to be said and done.

Subject: **Re: awol**
To: "Mcstotts, Roney L SFC NG NG NGB" <roney.mcstotts@us.army.mil>

Date: 07/07/10 11:58
From:

SFC Mcstotts;

I'm not worried about physical harm, I'm worried about being in a unit where the leadership puts getting someone in trouble over the truth and ignores a member when thry try to get something fixed.

----- Original Message -----
From: "Mcstotts, Roney L SFC NG NG NGB" <roney.mcstotts@us.Army.mil>
Date: Tuesday, July 6, 2010 22:39
Subject: awol
To: "Mahan, Wesley" <wesley.mahan@us.army.mil>
Cc: "carlos.enriquez" <carlos.enriquez@us.army.mil>

> SGT Mahan
> If you are suppose to be at drill and you are not there you are
> awol. You can not excuse yourself frrom drill. To be excused, you
> sate to me you need to be excused, if its ligit I will pass it up
> the chain. Job conflict, school conflict or family matters that
> have to be taken care of. Being paranoid is not one of them.
> No one is going to shoot you, or physcial attack you.
> Capt.Enriquez has ask you to contact him about your complaints,
> you have not the last I heard.
> Be aware I am haveing diffculties geting on line it takes me
> about 40 minuntes to get to ako and then I might get kicked of
> after 5 minutes
>
> SFC Roney McStotts
> S-6 NCOIC
> 1-108TH CAV SQDN

It was not a drill weekend. It was a voluntary retreat. Over 1/3 of the 1-108th did not go. I was the only one punished. I was punished after I had went to strong bonds to make it up. We just had to go to the event, it was held several different times.

Don't tell them what they have done tell them what they think they have done. I do not care how good it made joey recker feel to tell me I was not a real man because the girl got married (in the states) while I was in afghanistan, it was unprofessional for joey recker to do so. How does it sound to tell someone who was dumped while on an overseas deployment that it was their fault? If you know it is a trap it is not a trap.

Page 1 of 1

ect: drill
To: "Mcstotts, Roney L SFC NG NG NGB" <roney.mcstotts@us.army.mil>
Cc: carlos.enriquez@us.army.mil

Date: 07/12/10 11:04
From:

SFC Mcstotts;

Would it be possible for a jag lawyer to be at drill? If I am going to have a false official statement thrown at me I would like to have legal representation. I would also like to discuss retaliatory punishment and forgery.

Sgt Wesley Mahan

https://webmail2.us.army.mil/iwc_static/layout/shell.html?lang=en-us&15.01_143243 10/11/2010

I should have sent that one to more people. Direct and concise. They did not want that. When I went on my own to talk to a jag lawyer, he laughed at the situation but told me it was dangerous. If they will forge documents and signatures, they will do anything. They were doing enough to railroad me but not enough to involve the legal side of things.

Oh Lord why did they lie so much to convince themselves I was afraid of one of them physically attacking me? I mean I know why because that was the last physiological barrier, they could put up about the crimes they allowed and encouraged. But why? Why would leaders in an organization ignore crime and say the whistle blower is afraid of someone there attacking him?

As mentioned, before I once wrote the commander of the Georgia National Guard directly. The letter was returned to me with a stamp on it. There was a letter enclosed but sadly I lost that letter, it was a thing of beauty though. Here is my letter:

Sir:

I apologize for allowing myself to be forced to contact you regarding these issues but I have exhausted all of the Georgia Guard's options. When I was in the 1-108th I discovered that some individuals had been forging my signature. When I used my chain of command I was threatened. When I blew the whistle Lt. Col. Randall responded by writing a false statement as did Captain Carlos Enriquez. Lt. Col. Simmons later wrote a second false statement in attempt to clarify what he said. I have spoken to jag, the ig and Lt. Col. Simmons himself but nobody is going to do anything about the crimes and cover up. The false statement Captain Enriquez wrote was used to demote me 19 days before I was honorably discharged. Lt. Col. Simmons has since admitted in an email exchange that he lied on the statements he wrote but that doesn't help me if the ig wont do anything. I have copies of everything I have mentioned and more. I am looking for what I've been looking for since the beginning, why was my signature forged and how many times did it happen? It would explain a lot of the pay and record issues I had when I was with the unit. I didn't ask anyone to lie to punish me because I reported crime but I've had to deal with it.

Thank you for your time

Wesley Mahan

RECEIVED

JAN 2 0 2012

OFFICE OF THE ADJUTANT GENERAL
GA DEPT OF DEFENSE

Seems pretty simple, notice the nice stamp on the corner? How polite of the Georgia National Guard to let me know they received word of crimes being committed within the Georgia National Guard and did nothing. Nothing. Would that knowledge not make recruiters sweat when talking recruits and families of recruits? I mean if document forgery, signature fraud and whistle blower retaliation are accepted and rewarded by the Georgia National Guard how can anyone be safe? Is that only in the Georgia National Guard or do other parts of the military protect criminals as well?

This all ties together into a reveal given to me during my last conversation with someone from brian kemp's office. He defined as "governor" by the aide just did not like to do anything with regards to the Georgia National Guard. Yes he was seen as the commander in chief of the Georgia National Guard but there "really was not a mechanism for him to do anything" and he "didn't really want to make one". So who is the Georgia National Guard accountable to? That really sounded like brian kemp just did not want to do anything. Such bravery and integrity from a man who would not step down from his position as secretary of state until the secretary of state's office confirmed his victory in the 2018 governor's election. It is also a little sad that brian kemp is ok with crime in the Georgia National Guard. I wonder if Vernon Jones, Stacy Abrams or Hunter Hill would be ok with that? I'm quite sure Spike Cohen would defiantly not be ok with it.

Since none of the FOIAs were answered with documents the FOIAs were about I decided to send in two more. What better way to end it?

This FOIA concerns a letter written in reaction to an inquiry from State of Georgia representative Chuck Payne in 2019 and a hearing that took place on February third 2011. The paperwork that was sent to me has the date as 3 feb 2011 signed by john f king.

This FOIA seeks all documents, emails or written statements about the letter sent to Chuck Payne's office in 2019 regarding the inquiry about forgery, fraud and randall simmons lying to Senator Chambliss and the hearing about me from February third, 2011, signed off by john king

He responded:

Mr. Mahan,

Attached is the agency's acknowledgement and final response to your most recent FOIA request

V/R

COL Bischoff

DEPARTMENT OF THE ARMY
GEORGIA ARMY NATIONAL GUARD
OFFICE OF THE STATE JUDGE ADVOCATE
1000 HALSEY AVENUE, BUILDING 447
MARIETTA GEORGIA 30060-4277

Office of the State Judge Advocate

June 21, 2021

Wesley Mahan
Via email to lgtdrm@gmail.com

Dear Mr. Mahan:

We are in receipt of your FOIA request dated June 19, 2021. We have assigned your request Agency Control Number FA-21-0013. Your request is duplicative in nature to your previous requests, specifically FA-20-0001, FA-20-0009, FA-21-0004, and FA-21-0005.

This is a final response to your request(s) for records related to the personnel actions taken against you in 2010-11 as well as all requests related to MG Randall Simmons's response to Senator Saxby Chambliss in response to your request for Congressional assistance.

We are unable to accept or process any additional or subsequent request for these records. If you submit an additional request for records related to the above matters, that request will not be processed and you will NOT receive a response from the Agency.

The records we had have been furnished to you. Any other records you have requested either do not exist or have been destroyed in accordance with applicable record retention standards. You may appeal this determination to the Secretary of the Army within 60 days from the date of this letter. This concludes our office's processing of your request(s).

If you have any further questions regarding your request please contact my office at (678.569.5644) or by e-mail (brian.s.bischoff2.mil@mail.mil).

Sincerely,

Digitally signed by BISCHOFF.BRIAN.SCOTT.1258927784
Date: 2021.06.21 13:23:09 -04'00'

BRIAN S. BISCHOFF
Colonel, Georgia National Guard
State Judge Advocate

The FOIA was about a letter written in 2019 and Bischoff goes immediately to 2010/2011 and does not even mention 2019. That does not show personal integrity within the Georgia National Guard JAG office. So, they threw away stuff from 2019 that quickly? According to some of his previous replies they should have held onto it or maybe bischoff just does not want to talk about the letter from 2019. The FOIA was about the letter from 2019 but it did mention the hearing I was not invited to. Obviously, that was a mistake on my part as that reply showed what bischoff thinks of law apparently. A jag lawyer for the Georgia National Guard that protects criminals. Classic. Of course, I responded. Yes, it is just a modified version of one they never answered, that was the point. He put his refusal in all BOLD letters so that means he really meant it. I suppose he was tired of making excuses for criminals.

This also showed me that a FOIA does not mean anything to the Georgia National Guard. On these FOIAs the basic response was to either focus on the least worst part of it and answer that in a non-answer kind of way. Other parts they would ignore or respond by sending in documents that had nothing to do with the question. But hey at least now it is a known fact that the Georgia National Guard thinks little of the FOIA process. I wonder if bischoff really thought I would just lay down? At the very least he now knows how to spell my last name correctly. His reply listed five FOIAs, I only received three actual replies. Integrity of the officer core at its best.

Once again, the jag lawyer uses the term "responses". He is correct in that however there were never any answers or truthful information in those responses.

On July 8th of 2010 Randall Simmons wrote a letter to Sen. Chambliss about me. He told the Senator I had been involved with an "ivy ferguson". He repeated that claim in a second letter to the Senator which was dated May 15th, 2011. In an email dated july 9th 2011 Randall Simmons admitted that there was no "Ivy ferguson". The name "ivy ferguson" was thrown at me in Afghanistan as well as after we returned. This FOIA seeks all information that went into randall simmons telling the Senator that I was having an affair with a female that did not exist.

His reply, he did not disappoint me.

Mr. Mahan,

We have responded to your multiple requests for records on this matter. As per our last response, we are no longer processing any requests from you for records related to this or related incidents.

Respectfully,

COL Bischoff

I like it when someone ends a communication with "respectfully" when it obvious that person respects nobody. So now the response is to put their hands over their ears and scream "we can't hear you". What did he really think was going to happen? At least bischoff the jag lawyer finally spelled it out on how little he thinks of the law. How hard is it to tell me why an officer lied about me to a senator and violated my due process rights and why the Georgia National Guard let the officer get away with it? State of Georgia representative Chuck Payne would probably make brian bischoff or someone else tell me why they did not respond to the inquiry from State of Georgia representative Chuck Payne. Well, they did respond but they did not actually answer any of the questions, so I sent this:

Mr. Payne;

I have attached the reply the Georgia National Guard sent me in response to your inquiry in 2019. As well as a copy/paste of the inquiry I sent to you. I have contacted the Georgia National Guard several times on my own regarding the reply. They have refused to say why the response did not even mention the actual questions and instead mentioned something that did not affect me and I never asked about. It has regressed to them threatening me, telling lies about what has been sent and telling me they will never answer those questions. However at this point the only thing I need to know is why they replied to your inquiry with a letter that had nothing to do with the inquiry. All that matters is the Georgia Guard telling me, in writing, why they refused to answer the questions from you and replied with a letter that had nothing to do with the questions. The POC for this is Brian Bischoff, State Judge Advocate, 678-569-5644 or brian.s.bischoff2.mil@mail.mil

Thank you

Wes mahan

Amazingly simple but how would brian bischoff respond if he responded at all? Well, it turned out he would not but the governor of Georgia and commander in chief of the Georgia National Guard finally did. What would he say? Would he announce an investigation? Would he apologize for his office literally stamping return to sender on an envelope that contained incriminating evidence on members of the Georgia National Guard? Would some offer be made to somehow make up for what the Georgia National Guard did to me? I do not know if this was a result of Payne, Ossoff or myself. Also, brian kemp's reply was a big letdown.

STATE OF GEORGIA

OFFICE OF THE GOVERNOR

ATLANTA 30334-0900

July 26, 2021

Dear Mr. Mahan:

Thank you for writing my office to request assistance. I appreciate your concerns.

I have referred your information to the Georgia National Guard and asking that the appropriate staff member review the situation and respond to you. In addition, you may contact this office directly at:

Georgia National Guard

678 569-6060

I hope this information is helpful to you. Thank you again for writing.

Sincerely,

Governor Brian P. Kemp

So, the answer from Brian Kemp, governor of Georgia and commander in chief of the Georgia National Guard, is to refer the issue back to the department that started, allowed, and then refused to do anything about it. I wonder if all the information I sent to brian kemp's office had anything to do with brian kemp not wanting to do anything about it? I never heard from the Georgia Army National Guard or Georgia National Guard again, they were thrilled to not be hearing from me anymore. I did call the office of brian kemp back and inquired when I would hear from the Georgia National Guard. I got a repeat of a previous reply, Brian Kemp just does not want to be involved in criminal issues within the Georgia National Guard. I was told that he just did not want to make the effort, it was too hard to stop criminal activity in the Georgia National Guard.

It was easier to just kick the issue back to the Georgia National Guard under the guise of doing something. Pointless of course as the Georgia National Guard had performed Olympic level gymnastics to avoid answering any questions about or investigating criminal actions by high-ranking individuals. Even to the point that brian kemp, the commander in chief of the Georgia National Guard, refused to do anything. Now that is a politician with no integrity.

Chapter 16: Dead to Me

Every situation mentioned in this book could have played out differently. A service member is sexually harassed, perhaps even more, and the service member reports it as they should. The sexual predator that had been hiding behind rank is punished and hopefully the victim is given some counseling and realizes that the entire military is not full of sexual predators. Sadly, that did not happen in this case. The victim was attacked and made to feel shame for reporting a sexual crime. The victim was told, unofficially or otherwise, that the rank of the criminal made it ok for the criminal to commit crimes and that because of her rank she was supposed to be a compliant victim.

Now during my time in the military, I observed people of rank grooming lower ranks for sexual pleasure. Sometimes the lower rank was fine with it and sometimes the lower rank was unaware of it until it was too late. Now what should have happened in that case? The sexual criminal should have been punished, that did happen. However, the victim was not given the proper support and was instead punished for reporting the crime. That is an easy fix. If you attack a victim of sexual assault, you just ended your own career.

The forgery and fraud that blossomed into its own creation is a unique event. The solution is simple. Administrative sections should not be allowed to forge documents and signatures. That would have stopped everything in its tracks, but the Georgia National Guard has its own version of what is defined as legal. But beyond that how could each step have been stopped? First off, the forgery and fraud done by administrative sections of the Georgia National Guard needs to stop.

I do not believe that it is "only" the 1-108th that does that. The ease at which the inspector general office of the Georgia National Guard told me not to worry about it confirmed that as well as interactions with members of other units of the Georgia National Guard. It does not take a very observant person to look at some members and know that the height, weight, and physical fitness standards are not being enforced which in turn means that the paper test results are being forged. From there it is a simple logical step to know that the constant paperwork errors members have are not just a lot of little mistakes but in fact the documents they turned in being destroyed and forgeries being sent in with forged signatures.

So that needs to be changed. That change would also tie into changing the culture that allowed brian green to threaten a whistle blower and brag about wanting to get the whistle blower in trouble. How are those things elevated to the acceptance level in the Georgia National Guard? Does the Georgia National Guards internal leadership training need to be changed or is there an unofficial rule that states that high ranking people are not accountable to the law? That is a sad theory. Theory or not it needs to be stopped.

The straightforward way would be to have the law enforced. Forging documents and signatures should not result in a promotion and an award but punishment and jail time. Threats to subordinates reporting crime would be met with shift action not the IG refusing to do anything. Also, the process of how personnel get into the Inspector General, JAG and command positions while having no integrity and in fact protecting criminal behavior would have to be examined and stopped. How does a JAG officer think it is ok to threaten someone reporting crime? The core values, if applied, would solve all these things.

There would be no NCOs and officers that protected sexual predators because they would have been weeded out. The same with JAG lawyers that protect criminals, they would be gone as well. Overall integrity and accountability would solve and prevent most of the problems mentioned in this book. However, that seems to have become a problem for people of rank in the military. Look at Fort Hood. I would be willing to put money on no officer ever getting in trouble for the crimes and deaths there.

That is the real problem. This is not about me it is about the people who had similar things happen to them but did not have an officer stupid enough to put the lies and crimes in writing. The ease at which that all happened suggested that everyone involved was extremely comfortable with crime. Sadly, that was not confined to the 1-108th during my time in the military, it was like the core values vanished from some people as soon as they got high rank.

How else does sexual misconduct, let alone other crimes, become normalized in the military? Once upon a time while playing video games in a friend's room in rhien main Germany we both got quite a surprise. There was a noise at his window and a female service member that we both knew climbed into his room. She steadied herself on the floor looked at us and apologized.

She had been at The Rocket, the club on that base and had attracted the attention of someone did not want. When she left The Rocket, she noticed she was being followed. Followed in a way that was not safe as the "men" kept cat calling her on a military case in the early morning hours. Cut across a ball field no good and her dorm was a few blocks away but ours was not. So, she went to our building and seeing that the window as open climbed in, the "men" following her suddenly lost interest. If only all the criminals in the military were that cowardly.

The first time the air force sent me to Germany the night we arrived at rhien main half the married people on our trip went to the red-light district of Frankfurt or found another way to cheat on the spouse they had just left. Now adultery was in crime in the military when I was in, now I do not know but it is not about morality. It is the legality of the situation and the sad fact that most of those folks were of the middle to high rank and now they expect everyone else on the team to lie for them if asked.

Now that the lie for the ranking person has been established for one thing it is simple to apply it to another. However, the original lie must be defended with the zeal of a religion. The phrase "what goes on deployment stays on deployment" comes to mind. This does create a bit of a quandary. Mission critical things sometimes preclude things that the home station would not like but is the difference only in the mind? That is a little bit of a tangent. I do have a little bit of a story about that saying from my second trip to Germany.

This was with a different group of people a few weeks before the trip the enlisted person I shall call Sergeant cheats on his wife came up to me and said, "you know what goes tdy stays tdy". Now he had no idea I had been to Germany before and had also done a short trip at Dover AFB. He also had no idea how wild I was in my personal life. My answer displeased him. This all resulted in me being shown what I would vow to never become.

So, when we get to Germany it "gets back" to me that sergeant cheats on his wife had told all his friends on that trip that I would tell on them if they did anything illegal or unethical. This was all because I told him that only people who cheated on their spouses said his little mantra. I had no interest in their marriages nor in going out and getting drunk every night we were there so that made me bad. That was the problem. Because I would not validate their choices to cheat on their spouses and show up for duty drunk, I was seen as the problem.

The military has developed a culture wherein service members are expected to lie to protect people of rank when they commit crimes. When people rightfully object to that they are insulted and attacked. Rank covers for rank and the people that know it is wrong either get corrupted or get out. After that it is just a simple matter of constantly being dishonest to any elected official that asks about the crimes. Simple and effective, not good but it is what it is because nobody has the integrity and courage to change it.

You will not hear a recruiter talk about this. You will not hear an officer blowing smoke at a congressional committee talk about it either. It will not be mentioned at promotion and other military ceremonies, but it is there. Why does this not get dealt with? Core values as they are presented would preclude certain thing as not happening so where is the disconnect? Until the disconnect is found and stopped there will still be people of rank who abuse people for their own amusement and are protected by the military justice system. This must stop.

The United States Army War College

The United States Army War College educates and develops leaders for service at the strategic level while advancing knowledge in the global application of Landpower.

The purpose of the United States Army War College is to produce graduates who are skilled critical thinkers and complex problem solvers. Concurrently, it is our duty to the U.S. Army to also act as a "think factory" for commanders and civilian leaders at the strategic level worldwide and routinely engage in discourse and debate concerning the role of ground forces in achieving national security objectives.

The Strategic Studies Institute publishes national security and strategic research and analysis to influence policy debate and bridge the gap between military and academia.

The Center for Strategic Leadership and Development contributes to the education of world class senior leaders, develops expert knowledge, and provides solutions to strategic Army issues affecting the national security community.

The Peacekeeping and Stability Operations Institute provides subject matter expertise, technical review, and writing expertise to agencies that develop stability operations concepts and doctrines.

U.S. Army War College
SLDR
Senior Leader Development and Resiliency

The Senior Leader Development and Resiliency program supports the United States Army War College's lines of effort to educate strategic leaders and provide well-being education and support by developing self-awareness through leader feedback and leader resiliency.

The School of Strategic Landpower develops strategic leaders by providing a strong foundation of wisdom grounded in mastery of the profession of arms, and by serving as a crucible for educating future leaders in the analysis, evaluation, and refinement of professional expertise in war, strategy, operations, national security, resource management, and responsible command.

The U.S. Army Heritage and Education Center acquires, conserves, and exhibits historical materials for use to support the U.S. Army, educate an international audience, and honor Soldiers—past and present.

STRATEGIC STUDIES INSTITUTE

The Strategic Studies Institute (SSI) is part of the U.S. Army War College and is the strategic-level study agent for issues related to national security and military strategy with emphasis on geostrategic analysis.

The mission of SSI is to use independent analysis to conduct strategic studies that develop policy recommendations on:

- Strategy, planning, and policy for joint and combined employment of military forces;
- Regional strategic appraisals;
- The nature of land warfare;
- Matters affecting the Army's future;
- The concepts, philosophy, and theory of strategy; and,
- Other issues of importance to the leadership of the Army.

Studies produced by civilian and military analysts concern topics having strategic implications for the Army, the Department of Defense, and the larger national security community.

In addition to its studies, SSI publishes special reports on topics of special or immediate interest. These include edited proceedings of conferences and topically oriented roundtables, expanded trip reports, and quick-reaction responses to senior Army leaders.

The Institute provides a valuable analytical capability within the Army to address strategic and other issues in support of Army participation in national security policy formulation.